P9-CMN-669

For Kiyo, for everything

Contents

Acknowledgments

For an ethnographer, there are several forms of isolation to contend with in the process of producing a book such as this one. The immersion of fieldwork, though it involves a great deal of social interaction, is at the same time an experience of leaving the comforts of home and of a cultural world one understands with a natural ease and entering into a confusing, intriguing, and often uncomfortable new world. The analysis and writing of the ethnography itself entails hundreds of hours of silent contemplation, alone and behind closed doors. As a consequence, the moments of connection and support along the way are that much sweeter. I have many people to thank for their professional and personal assistance.

My initial fieldwork was greatly assisted by the Opuni family, particularly the introductions facilitated by Emmanuel Opuni, a fellow student at the London School of Economics. Stephen and Comfort Opuni adopted me as a daughter though I was a total stranger to them and put me up in their home in East Legon for over a month. When I found a place to live in town that was closer to my fieldsites, demonstrating the matter-of-fact generosity and patience that I have come to associate with Ghanaian hospitality, they wondered what my rush was to leave. I must also thank the crew at BusyInternet, in particular Mark Davies and Estelle Akofio-Sowah, who were both enormously helpful. People who helped me learn the basics of the Twi language include Isaac Okwaning and my Twi instructor at the University of Ghana-Legon, Professor O. N. Adu-Gyamfi. Deputy Commissioner Isaac O. Apronti of the Customs, Excise and Preventative Service of the Government of Ghana was enormously helpful in providing the data on computer imports referenced in chapter 7. Ross Barney and John Burg introduced me around and helped to ensure my downtime in Accra was fun. Jon Roberts, a fellow researcher and friend, aided with contacts, references, and suggestions over the years. Other researchers in Ghana whose company I was pleased to keep at different stages of this project include Richard Stanley, Matti Kohonen, Bianca Murillo, and Denise Nepveux.

As a graduate student in the sociology department at LSE, where I wrote my PhD dissertation that formed the basis for this book, I must thank my dearest friend, Jill Stuart, a fellow Oregonian, who met me for cappuccino, croissant, and conversation every morning at school. Faculty at LSE and elsewhere who graciously agreed to serve on my committees and who thoughtfully reviewed my work with extensive and insightful feedback include Fran Tonkiss, Sarah Franklin, Lucy Suchman, and Daniel Miller. My thesis advisor, Don Slater, shepherded this project from its earliest stages. My work is greatly, greatly improved by their contributions. Mentors from my time working at Intel include Ken Anderson, Genevieve Bell, John Sherry, and Christine Riley. At Cornell University, without Geri Gay I would not have become a researcher at all.

At UC Berkeley I wish to thank especially Abena Osseo-Asare, whose support and sharing of similar experiences came at a critical time. Cori Hayden and some members of the Science, Technology, and Society Center read an early version of the introduction and offered thoughtful comments. The African Studies Center at Berkeley has been a great community to be involved in and much credit must be given to Martha Saavedra, the center's indomitable director. Students at the School of Information who were thoughtful and engaged readers include Dan Perkel and Heather Ford. I am also pleased to have such good colleagues there, in particular, Coye Cheshire, Tapan Parikh, and our Dean AnnaLee Saxenian. A big thanks to Julie Cheshire for working magic on the photos that appear in this book.

Some materials in this book have appeared elsewhere in other forms. Chapter 3 originated as a journal article, "Problematic Empowerment: West African Internet Scams as Strategic Misrepresentation," *Information Technology and International Development* 4.4 (2008): 15–30. Likewise, chapter 4 originated as an article, "User Agency in the Middle Range: Rumors and the Reinvention of the Internet in Accra, Ghana," *Science, Technology, & Human Values* 36.2 (2011): 139–159. I wish to thank my editor at the MIT Press, Marguerite Avery, and the Acting with Technology series editors, especially Bonnie Nardi, whose encouragement helped buoy me through the final stages of writing and revising. Three anonymous reviewers were

thorough, thoughtful, fair, and got reviews on the manuscript back to me remarkably quickly. I'm very grateful for their generosity.

My family gets much of the credit for their guidance and support throughout including my parents, Allen Burrell and Peggy Baldwin; my sister, Rebecca Burrell; and my extended family Rae Baldwin, Steven Baldwin, and Grace and Wilfred Burrell. The most directly affected in all of this is my husband, Kiyo Kubo, who demonstrated superhuman patience while maintaining our relationship long distance for three years as I lived first in London and then in Ghana. His enthusiasm for sharing everything with me from the mundanity of domestic life to discussions of esoteric scholarly topics makes him my perfect partner in life.

I've reserved the very last spot in the acknowledgments to thank my research assistant, Kennedy "Kobby" Ansah Asare, who made several critical parts of this book possible. He has become my most trusted ally and closest friend in Ghana and makes every return visit a delightful reunion with a place that has become like my second home.

1 Introduction

All technologies incite around them that whirlwind of new worlds. Far from primarily fulfilling a purpose, they start by exploring heterogeneous universes that nothing, up to that point, could have foreseen and behind which trail new functions.

—Bruno Latour, "Morality and Technology: The End of the Means," 250

Newlyweds Joyce, a twenty-four-year-old Ghanaian woman, and George, a fifty-one-year-old Canadian man, sat across from me at a pizza restaurant in mid-November 2004. From balcony seats, lit dimly by strings of Christmas lights, we relaxed into the interminable wait for our soggy pizzas. Below us lay the teeming streets of La Paz at dusk, a mixed middle- to low-income suburb of Accra, Ghana's coastal capital city. Street sellers piled their sour, green oranges and lined up mobile phone accessories on tables and tarps lit by kerosene lamps. Rigged sound systems drew attention to the wares of cassette tape sellers who pumped music so loudly the speakers visibly vibrated. *Tro-tro* drivers screeched confidently and recklessly between vehicles and pedestrians, the driver's *mate* hanging precariously from the window, rhythmically shouting his route. This was George's first visit to Accra and first time setting foot on the African continent that had no particular draw to him apart from this young woman whom he had just married.

George and Joyce had met five months prior on an Internet matchmaking Web site called plentyoffish.com. In our conversation they described how travel visa troubles thwarted their initial plan for Joyce to come to Canada to pursue the relationship in person. They believed that marrying would help Joyce acquire a visa and subsequently relocate, something George thought would likely happen within the year. Their modest nuptials had taken place a few days before. Joyce wore a white chiffon wedding dress that George had picked out and gingerly carried with him on his multileg journey from Ottawa to Accra. They carried out the civil ceremony and license signing in the presence of only a few friends and onlookers and a hired videographer.

Our face-to-face meeting offered some concrete proof (much to my skeptical surprise) of what young Internet café users in Accra had been saying all along about how the Internet really works. Joyce described a personal history that, similar to so many young Ghanaians, was marked by the struggle to keep the momentum of her education going. She had migrated toward urban opportunity a few years back with a boyfriend from her rural hometown. He had promised to help fund her education. He was a similarly aged young man, whose jealousy eventually led to their breakup, though he did make good on paying for her secretarial training. Turning away from such "young boys," she shifted her preference to older, foreign men whom she characterized as more loving and less stingy. The Internet offered a new way to meet men who fit this description through its numerous dating sites. For his part George still had questions about their fast-moving relationship and about her attraction to him. He appeared taken aback by the young, high-spirited woman he had just wed who didn't talk as much as giggle in his direction. Joyce admitted that she was reluctant to inform her extended family of the marriage for fear of judgment and requests for support, revealing some of her unease about the relationship in terms of its social appearance. As some indication of the validity of Joyce's concerns, when I relayed the story to the young Ghanaian mother of the family I was staying with, she bluntly reduced Joyce's motives to one word: *money*.

The initial, fateful meeting of George and Joyce transpired, in a sense, at a place only a short walk away. If one were to descend from the pizza restaurant into the streets, across the main paved highway, and down wide, dusty, potholed roads, past walled compound houses into a darker, quieter part of La Paz, one would arrive approximately ten minutes later at a small Internet café of the sort typical for Accra. The owner had named it *Sky Harbour* after the Phoenix, Arizona, airport, the city where his sons were living. It was situated on the ground floor of a two-story concrete brick building in a bare but air-conditioned room. The room held ten computer workstations, each separated by wooden privacy partitions. Though such small businesses referred to themselves as Internet cafés this was generally a misnomer because no food or drink was served.

Throughout the day activity levels at Sky Harbour waxed and waned. At midday it was often still and silent, with one or two users sitting attentively before the computer screen, often deaf to any surrounding activity. Deeply immersed in online social experiences, their physical presence in the café became muted and hollow. Such users could be carrying on conversations with someone who might be almost anywhere in the world: Indonesia, Palestine, China, Poland, or the United States. In the afternoons, groups of schoolmates sometimes wandered through to view American hip-hop and rap music videos. From the evenings until late into the night activity shifted toward a relaxed, lively, even mildly sociable atmosphere. Bright fluorescent lights illuminated the digital oasis and very loud music played, sometimes from the chop bar upstairs or from the machines themselves with one song layered incongruously over another. In her daily chats with George, Joyce had been one of the very purposeful and focused users in this space, her eyes trained on the scrolling chat conversation and the Webcam image of her beloved, indifferent to her physical surroundings.

At my first meeting with Joyce and George, Joyce seemed poised to realize an idealized vision of the Internet and its promise that was pursued by so many youth in Accra's Internet cafés. Her experience points to the more general shared view among these youth of the Internet as a mechanism for expanding one's social network, specifically in order to acquire foreign contacts. The desired outcome was a profound alteration or acceleration of one's life course with support from these well-situated outsiders. In Joyce's case, she had formed a bond online that was so powerful it compelled her foreign contact to make a physical appearance in Ghana and unlocked a flow of resources in her direction. The course their relationship had taken held the very real promise of liberating her from her state of geographic boundedness transporting her *aburokyire* (meaning *abroad* in Twi or more literally *beyond horizon*) as well as elevating her (through marriage) from a prolonged youth into respectable adult status. The image of fishing evoked by the plentyoffish.com matchmaking site's name is apt. Fishing in an abstract sense leverages the barest thread of connection (similar to a low-bandwidth digital channel) to hook something from the vast sea of possibilities. It also evokes something of the practice of phishing, the form of Internet fraud that involves persuading random contacts to give up passwords and financial and identity information. Joyce saw the applicability of the metaphor in her match with George, as she commented: "I thank God that I got my fish." From repeated conversations and encounters with Joyce during the course of five years, I do not doubt that there was a genuine bond in her relationship with George. Yet there was also an inkling of the potential fissures in their relationship in Joyce's rushed pursuit of George and the role he was coming to play as the primary person on whom she projected the full weight of her aspirations.

George noted that Joyce had proposed marriage to him very early on but her seeming impatience ought to be put into its material context. Paying to use the Internet per minute or hour is expensive in Accra and such expenses quickly accumulate over time. It was George who had insisted on the five-month period of getting to know one another because, at first, he was "still pretty clueless as to what her motivation might be." Even after marrying Joyce, he still seemed not to have resolved this question satisfactorily for himself noting, "I still don't know why; I tried to find out and you don't get a good answer. She just liked the look of me I guess." Their relationship had developed a kind of intimacy in five months of constant chatting online though given the physical distance between them they had only an abstract sense of what a life together would be like . . . Despite this, they shared a sense of optimism and a mutual reluctance to probe too deeply into any dark corners of doubt.

This book considers nonelite, urban youth such as Joyce and their changing sense of the wider world and their place within it as formulated through their involvement in Accra's Internet café scene. Specifically, the Internet became a key resource for enacting a more cosmopolitan self. With the rapid acquisition of foreign ties, the geographic scope of operation for these youth began to expand. They navigated this new social terrain in a number of roles—as individuals, as members of various social groups, as citizens of the nation-state, and in transnational or diasporic spheres. The anthropologist Arjun Appadurai argues for *imagination* as "a constitutive feature of modern subjectivity" (Appadurai 1996, 3), noting how narratives and imagery circulating in the electronic mass media are tied to aspiration and agency (Appadurai 1996). One can see that the Internet cafés offered a whole new realm of mediated exposures for youth. Yet the Internet in

Accra did not serve simply as a resource of imagery and ideas for these youth, a stimulus for activities carried out elsewhere in the mode of Appadurai's work. Rather it was a deeply interactive space in its own right to be acted on and through. In the digital traces and encounters of users on the Internet, the work of imagination is critical social practice but in the context of the very real possibility of failure and disappointment—the pushback of engaged materiality. Joyce's success story belies the more miserable record of many Internet café users in Accra in successfully forming and leveraging partnerships with foreign contacts. Youth in Accra's Internet cafés frequently struggled to decode the implicit social norms in virtual online spaces (derived from Euro-American codes of social interaction) and were often met with silence, avoidance, and exclusion when they fell short.

Some specific changes in the capabilities and practices of youth and the broader social order that emerged from the Internet cafés are worth noting. First, the public space of the Internet café in Accra offered a new capacity to orchestrate encounters across distance and to accumulate foreign ties. This capacity, once limited to the elite, university-going, transnational class in Ghana, was extended to a new segment of the youth population through the relatively accessible café space. These young Internet users were decidedly nonelite, marginally employed, and had a degree of education that they had struggled to obtain and subsequently struggled to leverage. Tellingly, the youth who were particularly drawn to the Internet café were often those for whom other avenues of similar enactment (particularly migrating abroad) were blocked. A second area of change emerged from the way these users contended with the spatial ambiguities of the online environment. In chat room conversations and online profiles, these youth constructed an alternate online self through text and imagery seizing on the possibilities of altering race, gender, age, and location. Although previous research has linked this capacity in virtual spaces to the self-focused identity exploration of users in the West, in Ghana it was recognized and leveraged in a very different mode to perform for and persuade foreign others and at the furthest extreme to carry out Internet scams. Finally, there was also a change taking place in the extension of religious cosmologies to the online realm. For some religious leaders and faithful followers, the enhancements of the Internet generated a new sense of supernatural force as traversing electronic links to encompass sites and subjects globally. Formerly distant outsiders came to be placed in relational roles and subjected to forces once reserved for kin. This was shaped, in part, by the distance-defying sense of co-presence and the immediacy of feedback online.

Interpreting Technology in the Peripheries

The chapters that follow offer a richly descriptive cultural account of how the Internet came to be distinctively materialized in urban Ghana and elaborate on the claims made in the previous section. This account demonstrates in numerous ways the interpretive flexibility of technology, that the meanings and uses of a machine or system are not predetermined by the form alone but come to be understood in distinctive ways by different user populations and other relevant groups. In taking this stance, this book joins many other works of ethnography and historiography (Miller and Slater 2000; Lally 2002; Horst and Miller 2006; Fischer 1992; Bijker 1995). These prior works treat a range of social circumstances as sources for the diverse interpretations of technology. The structuring of gender relations and gender identities has been especially well considered (Fischer 1992; Bijker 1995; Kline and Pinch 1996; Cockburn and Ormrod 1993). Additionally, national identity (Miller and Slater 2000; Nye 1996), accepted practices of social and economic obligation (Horst and Miller 2006; Donner 2007), and the cultivation of domestic life (Lally 2002; Silverstone, Hirsch, and Morley 1992; Shove 2003) are all offered as interpretive frames. By and large these works consider the negotiation of new technology taking place within a continuous (if contested) "cultural zone"[1] (Oudshoorn and Pinch 2008, 549). However, a move to the fringes, into spaces of cross-cultural encounter in particular and where the more erratic processes of globalization are operative (as considered in this book), means confronting discontinuity and some odd and novel surprises in technology's circulation and interpretation. Once brought into the conversation, such discontinuities stretch and deform prevailing analytical models of the relationship that connects users, developers, and technological artifacts and in this way have the potential to contribute new insights to the literature.

The culture of Internet use in urban Ghana developed from historical Global North–South relations and inequalities. It was shaped by the country's peripheral political and economic standing as experienced in the day-to-day lives of its citizens. The legacies of colonialism and subsequent independence movements across the continent, the rise of global governance and the international aid industry, and increasing exposure to the global economy through the neoliberal reforms of the 1980s and 1990s have all contributed to forming Ghanaian national and ethnic self-conceptions and modes of acting. It shaped the yearning for foreign contacts, travel experiences, and an entrenched notion of success as related to one's ties to abroad and to one's global mobility, which guided Ghanaian youth in their online forays. What I observed in Accra's Internet cafés suggests that the experience of the enduring marginality of these users, as related to these historical trajectories, is an additional interpretive frame. This marginality can be defined as a position of relational disadvantage, of noncentrality but not absolute exclusion. It was reflected in the power dynamics between young Ghanaians and foreign others online. To consider such asymmetries within a state of connectivity is to move beyond the digital divide as the defining concept in conversations around digital inequality. The metaphor of the divide is framed around a binary of access and nonaccess and an absolute distinction between users and nonusers. Anthropologist Brian Larkin in his study of media and technological infrastructures in Nigeria suggests that what makes this discourse problematic is that it "looks through rather than at the object at hand" (Larkin 2008, 235). I further assert that in light of the spread of global connectivity, the tendency to consider populations in Ghana and other economic peripheries principally as nonusers is becoming more and more of an analytical dead end.

What the Internet has become in Ghana stems in part from a particular global ordering and the situated perspective of young, urban Ghanaians within it. It is not a product of something as static and self-contained as what might be referred to as the Traditional Culture[2] of Ghana or of its many ethnic groups. I wish to clarify briefly how culture is instead understood in this book. Appadurai suggests that despite the ambiguities of the term there is value in retaining culture "as a heuristic device that we can use to talk about difference" (1996, 13). What must be resisted, he adds, is the notion of culture as a substance or "property of individuals and groups" and in this way as bounded, naturalized, and depoliticized (1996, 13). Along these lines I wish to refocus on a notion of culture as constituted in practices that come to be distinctively patterned and situated in place and time but that may feed on and integrate a multitude of influences and resources including those that are distant in origins or global in scope. This is not a new definition of culture but rather aligns with performative and practice-oriented definitions consistent with recent directions in the science and technology studies (STS) literature[3] that heavily informs this book. Such an understanding of culture as practice also best accommodates the potential for change through the initiation of its members. What I add to this is a particular sensibility about culture as dynamic, syncretic, and unbounded as reflected in some of the writing on globalization (Hannerz 1987; Appadurai 1996; Hannerz 1992; Lee and LiPuma 2002). Culture is forged in the ongoing push and pull between competing cultural threads and in the unique hybridization of forms.

By titling this book *Invisible Users* I have selected a multivalent term to characterize the experiences of this marginality in particular among Ghanaian youth who inhabited Accra's Internet cafés. On the one hand, their invisibility follows from this noncentrality as defined by a lack of accommodation by those in power (ranging from distant technology designers and network administrators to local politicians and other authorities). On the other hand, invisibility may be conceived of as its own form of power allowing forms of transgressive behavior to go unchecked. In this mode, law scholar Lawrence Lessig speaks explicitly about invisibility in conjunction with the freedom of the early Internet where users had greater anonymity. Users who were not identifiable were thus beyond regulation. As Lessig notes, this has changed over time with the encoding of new layers of traceability into the network infrastructure (Lessig 2006). The consequences of such a transition toward visibility and the implications for Ghanaian youth and their forays on the Internet are revisited in this book's conclusion.

One form of invisibility with this particular group of Internet café users follows from their uptake of the technology outside of formal design, manufacturing, and distribution processes. Here it should be noted that the Internet cafés of Accra were primarily equipped with secondhand computers that came to be distributed via gray market processes.[4] These machines were imported from abroad typically through the efforts of Ghanaian transnational small businesses. A lack of influence on the production end of the process limited the adaptive practices of Ghanaian youth to post hoc efforts to make up for the gaps between technological possibilities and local needs, realities, and interests by leveraging unintentional and serendipitous compatibilities.

The invisibility of this user population is also apparent in the communication processes taking place around and through the Internet. Young Ghanaian Internet users in Accra's cafés find themselves negotiating representations of Africa and Africans in their online encounters with foreigners. Ghanaians experienced a particular form of invisibility in virtual online spaces through their struggle to be seen beyond Western mass media caricatures of Africa. For example, Internet scammers found that adopting and manipulating Western representations of Africa by representing themselves as a corrupt politician or a needy African orphan was an avenue to financial gain. A struggle over representation is similarly underway in the international aid sector where the recent efforts by institutions such as the United Nations (UN) and the World Bank to claim digital technology for socioeconomic development practices has often meant denying the existence of competent technology users in "developing" countries such as Ghana. Such a discussion has remained anchored to the digital divide's simplified dichotomy between users and nonusers. Africa is thus depicted as a blank slate where technology has yet to arrive or, as one article described it, "a technological desert" ripe for intervention (Odedra et al. 1993, 25).[5] This is another process that renders these users invisible.

A larger story here is the way Africa and Africans are continually invoked as a metaphor in Western modernist projects—specifically, as a metaphor for deprivation, breakdown, or absence (Ferguson 2006; Mbembe 2001) or alternately for the authentic (Ebron 2002), closeness to nature, and a romanticized communalism. This work in metaphors emerges once again in more recent discussions of Africa's place in the digital age. For example, it is apparent where Manuel Castells situates Africa's significance in the global order as the place of humankind's origins, "the land that nurtured the birth of Lucy," but whose economic and social collapse "denies humanity to African people, as well as to all of us in our inner selves" (Castells 1998, 82–83). A developmentalist path to progress is apparent in such depictions that contrast the natural state of humankind's origins (as disordered, sensual, and timeless) with the advanced progress of Great Nations.

I seek to replace such metaphors with an account of one small corner of the African continent as a real place, one that is complex, changing, and unique and that can be more richly understood through a stronger commitment to the empirical and to the specific grounding offered by fieldwork. Work within the African studies literature has rarely been brought together with STS. The following section considers the unexplored possibilities in leveraging the accumulated base of such regionally specific knowledge and insight. Among other issues, this literature confronts the various ways Africa has become a projection of Western fantasy and what the consequences are for Africans in the real world. These projections now also extend into the virtual worlds inhabited by these invisible users such as the youth who frequent Ghana's Internet cafés.

Weak and Strong Materiality in Cultural Accounts

One principle aim of this book is to fill a void between scholarship on the global transitions and problems of inequality in the digital age and localized cultural accounts of the uptake of such technologies found in media studies, some parts of mainstream anthropology, and in occasional work in African studies. There is in these separate areas of scholarship an emerging and implicit opposition between two methodological and theoretical stances. One is best exemplified in the works of Manuel Castells, which depict the general weight of a worldwide shift as it is accompanied, accelerated, or initiated by some new material reality (Castells 1996, 1997, 2001; Castells et al. 2006). For Castells it is network structures in general and the Internet specifically that play a leading role in this transformation. In the writings of Negroponte (1995) and Mitchell (1996) the critical shift is from atoms to bits. An opposing viewpoint pursued using an ethnographic approach and explicitly contending with questions of culture points alternately to diverse and regionally particular patterns of media engagement (e.g., Ginsburg, Abu-Lughod, and Larkin 2002). This ethnographic work takes issue with totalizing claims of global transformation and points instead to the resilience of locality and of diverse systems of social value. Such work highlights locally peculiar trajectories of change in the wake of transformations to the media environment. Yet such assertions ultimately do not evaluate the claims of these digital age thinkers (which I maintain is still sorely needed) but instead embark from a separate premise. Such cultural accounts often implicitly adopt a kind of weak materiality that presumes certain limits on the consequentiality of objects and infrastructures. Technological form is subsumed under ostensibly more social projects of state building, class conflict, or forms of activism positioning the technology as a handy resource but not in any sense an initiating force. In this way, such scholarship opts out of the discussion of what precisely is *new* about new media technologies.

This book begins from an alternate starting point by insisting that some socially constitutive process necessarily occurs whenever populations engage with new material possibilities for any sustained period of time. That indeed "all technologies incite around them that whirlwind of new worlds" as Latour (2002, 250) notes in the epigraph at the beginning of this chapter. The question is what such a change looks like and how enduring and extensive it is in the particular setting in question. The repositioning I propose is greatly influenced by the work of anthropologist Daniel Miller, who in his decades-long critique of the "tyranny of the subject" in social anthropology (Miller 2005, 36) challenges the way material culture is relegated to the role of a passive, supporting player for an a priori social structure (Miller 1987). Following from this starting point, my intent is to devise a framework to guide the analysis of material forms and technologies of mediation in terms of their regular and specific qualities. Marshall McLuhan attempted this in his idiosyncratic consideration of the way light directionality, television screen resolution, and other technical properties can yield profound social reconfigurations (McLuhan 1964). Walter Benjamin does it as well in examining mechanical reproduction and the photographic arts that make possible images beyond ordinary visual perception. This reproduction, he suggests, destroys the aura of original art forms, reflecting and enabling a societal urge "to bring things 'closer' spatially and humanly" (Benjamin 2001 [1936], 52). McLuhan's fixed and immanent theory of materiality (where television screen characteristics seem to directly yield a societal shift toward communalism) is not precisely what I have in mind for this book but the principle of considering directly and at close range the technical and material as socially consequential is well worth keeping in mind.

In what follows I will consider some of the terrain of analysis so far left unexplored in cultural accounts of technology diffusion and appropriation among Africa specialists as they comment on colonial and postcolonial conditions. The specific works I refer to as *cultural accounts* are those that are historically and regionally well specified, often drawing on the richness and holism of ethnography and the accumulated work of

Africanist anthropology. By referring to the weak materiality of such cultural accounts my intent is not to judge such work as unscholarly (as in a weak argument). Instead *weak* should be read as a property or attribute (as in weak tea). It reflects a kind of self-imposed limit on how much sway a social scientist might admit to the material.

To take on a materialist stance (whether weak or strong) is to acknowledge that the consequentiality of objects in the social world in some way goes beyond what human intentions invest in them. Often it is a general durability and visibility of things (as opposed to ephemeral human interactions) that are seen as contributing to the making of society. Thus Douglas and Isherwood in their theoretical exploration of how goods compose a system of cultural expression note that "goods are most definitely not mere messages; they constitute the very system itself. Take them out of human intercourse and you have dismantled the whole thing" (Douglas and Isherwood 1979, 72). Law, Latour, and other elaborators of Actor-network theory (ANT) make this point by imagining the limitations of a purely somatic society perpetuated by bodies and their interactions alone (Law 2002; Latour 1991, 1992). As Law notes the material world is especially significant in sustaining the coherence and endurance of societies over time and across distances: "some materials last better than others. And some travel better than others. Voices don't last for long, and they don't travel very far. If social ordering depended on voices alone, it would be a very local affair" (Law 1994, 102). Some works of media theory also pick up on this, notably Anderson's work on imagined communities, which considers the role of print newspapers in connecting and sustaining the nation-state as an entity (Anderson 1983).

Existing scholarship reflects different degrees of commitment to a materialist stance. Analysis underpinned by a weak materiality acknowledges the significance of the durability and visibility of objects and yet dwells mostly in the realm of the symbolic. In this mode objects demonstrate one's status or communicate nonverbally some aspect of identity. For example, the sewing machine sitting as a kind of monument in front of the home of a tribal chief is put forward as an example of the prestige object's capacity to totally overcome and supplant its more mundane utility. At the heart of this formulation is an act of divorcing the aspect of an object available to sociocultural analysis from its (presumably nonsocial and material) properties and functions. This division goes back to Marx, whose notion of use-value referred to what was inherent to the commodity in isolation, the needs it served. Exchange-value, by contrast, was tied to a commodity's mediating role in society. Douglas and Isherwood employ this division between base functional uses versus cultural uses as their primary analytical angle stating, "Forget that commodities are good for eating, clothing, and shelter; forget their usefulness and try instead the idea that commodities are good for thinking" (Douglas and Isherwood 1979, 62). A further marker of a weak materiality is the nonspecificity or interchangeability of artifacts such that the case made for an object's role in social processes would be essentially the same if it were swapped with any other object. It is rather defined by its social history, where it has traveled, and with whom (Appadurai 1988). Theories of gift exchange exhibit this quality—it is the act of gifting (in relation to the sequence and nature of previous acts) irrespective of the particular thing gifted that strengthens social ties (Mauss 1990 [1950]; Bourdieu 1977). Such work is helpful in showing the broader ways that things are productively used in sustaining society beyond the more self-evident functions (e.g., that food is for eating), though they do not do much to challenge this naturalization of function as fixed and inherent to the material form.

Generally speaking, an analysis that validates the uses of commodities beyond apparent function has been valuable to the study of foreign commodities entering into colonial or postcolonial societies. Larkin refers to the way past African studies analyses of technology "stress how Africans understand and indigenize foreign technologies in their own conceptual schema" (Larkin 2008, 8–9). This type of work, he notes, has played an important role in countering colonial narratives of African incomprehension and the inability to "correctly" adapt to such alien artifacts. For example, Luise White's study of rumors about colonial vampires considers stories that employ fire trucks, hospitals, and other tools and practices as part of a symbolic repertoire composing a local commentary on the "extractions and invasions" (White 2000, 5) East Africans suffered under colonial rule. The acquisition and use of Western material artifacts or practices of wearing Western clothing styles has also been multiply interpreted. Anthropologist Jean Comaroff considers the partial embrace of Western dress evidenced by a Kwena chief in southern Africa who had a European-style suit made out of leopard skin. This act of combining symbols of chiefly office and Western authority she interprets as "a desire to harness the power of *sekgoa* [*white things*], yet evade white authority and discipline" (Comaroff 1996, 30). Ferguson, however, has argued against the notion that emulation is necessarily a form of resistance or subversion. He gives the example of one Sesotho man's expressed desire for a Western-style house with a steel roof and multiple rooms—something that in certain ways can be said to be unsound or "inappropriate" for the setting. This desire he reads as a "claim to a chance for transformed conditions of life—a place-in-the-world, a standard of living" (Ferguson 2006, 19). Such cultural accounts of how foreign commodities were received specifically in African urban and rural settings typically consider the valence of these impositions—whether they were in some sense left undigested constituting a kind of consent to domination or were rather redirected toward self-defined projects and sense-making among native Africans.

Work undertaken by Africanist scholars in particular has begun to carve out a niche examining especially breakdowns, maintenance, make-do, and bricolage as part of the equation of materiality (Verrips and Meyer 2001; Larkin 2008; Spitulnik 2002b). Verrips and Meyer (2001), for example, contrast the maintaining of cars by any and all available means in Ghana with the alienation of Westerners to their own cars, whose inner workings are mysterious and opaque and handed over to be managed by specialists. The Ghanaian jalopy continually projects itself not as singular whole but makes visible its true composition of many elements. This type of work on the constancy of breakdowns in everyday life is at odds with more mainstream STS in which breakdowns are portrayed as out of the ordinary, as sudden and transitory events.[6] Instead they are a fact of everyday life in places marked by scarcity and supplied with secondhand and overworn commodities. This raises the question of what behind-the-scenes upkeep enables a perception of material durability and to what extent this durability is itself illusory. This emerging work on breakdowns-as-everyday-life challenges a weak materiality that naturalizes the stability of material forms.

Brian Larkin must be credited for taking a concern with breakdowns the furthest as a defining concept in an African studies analysis of technology. As he documents, when British colonial powers in Nigeria built an electrical system or radio network they sought to employ it as a kind of spectacle to draw the association between colonial power and these particular awe-inspiring forms. When infrastructure breaks down, however, as long as such a symbolic projection remains in place, claims of power and progress are undermined. The radio network in its nonoperative state offers incompetence or disorganization instead. Larkin describes how, running counter to efforts at enrollment of those in power, "infrastructures have become the means to critique the state and lament the failed promises of elites" (Larkin 2008, 246–247). This engages a theme of recent social theory pertaining to materiality—the potential for material betrayal of artifacts deployed in social projects.[7] The betrayal Larkin notes is of a particular kind—linked to the way the material reality of an artifact or system is always unavoidably surplus to intentional manipulations of its symbolism. And yet this argument exhibits a kind of nonspecificity about form in dealing with the general entropic inevitability of things falling apart. One breakdown is no different than any other and has little to do with the regularities of the form, its specific properties, the distinctive mediation role or functionality of radio, or any other media infrastructure.

The reason for drawing a distinction between weak and strong materiality is ultimately to show how they operate quite differently in cultural critique. What a weak materiality offers principally is a mode of cultural reproduction via material adaptation. It is the malleability of form and its meanings and uses that are foregrounded. The previous examples demonstrate a push and pull between prevailing colonial or postcolonial institutions and the populations coping with or enduring such upheaval as it is rendered on a material medium. Yet the material is no more than a kind of firm surface that carries social forms unperturbed across time and space, a mute pawn in what is at its base a sociopolitical struggle. To get beyond a weak materiality we must consider instead a deeper entanglement whereby the novel constitution of the social takes place through the refractions of the material.[8] Although it is certainly the case that new, foreign commodities may be accommodated by a given cultural scheme, they are not simply reduced to this scheme. Larkin's consideration of breakdowns, of the material undermining symbolic projections, represents one view on the surplus spillover of the material. The strong materiality I argue for goes a step further to consider how the regularities of different forms become socially consequential.

In the quest to reconceive cultural accounts of technological engagement, I find analytical purchase in the notion of *relational materiality,* a term John Law employs to refer to a semiotics rendered in the object world

(Law 1999). Objects themselves are "effects of stable arrays or networks of relations" (Law 2002, 91). This means that rather than look for an object's immanent material properties (presuming its consistent, invariable material effect in the world), one looks for how this materiality is distinctively expressed in the way an object comes to be situated in unfolding actions. An object may thus be made material by what surrounds and engages it. In particular, it may be materialized by users (and other actors) who define novel ways of relating it to the other entities in their life world. The enduring value of a relational materiality to the pursuit of cultural accounts is the way of treating objects and specifically technologies as materially consequential but flexibly and nondeterministically. This materiality is not fixed, not arbitrary, not purely a "social construction," but situated and thus demanding in situ analysis.

Law's work on relational materiality is often grouped with the broader literature on ANT, which is most closely linked to the theoretical work of Bruno Latour. Latour has in the past argued provocatively and controversially for a symmetrical treatment of humans and nonhumans in their contributions to social processes. He makes a strong claim for the agency of nonhuman entities. This particular notion of agency is linked to the way an entity may "modify a state of affairs" (Latour 2005, 71). Latour's agency of nonhumans is without reference to intent and thus diverges from other writing on agency in which intentionality is core to the concept (Gell 1998; Ahearn 2001). I wish to take a few steps back from the analytical precipice of such rigorous symmetry between human and nonhuman, which is not, I find, essential to re-envisioning cultural accounts in a more materially grounded way and furthermore requires untenable sacrifices to accommodate. This notion of agency without intent creates a problem of explaining especially the initiation of action.[9] In considering technology adoption decisions that are wholly voluntary as in Accra's Internet cafés, this matter of initiation is a critical question. Furthermore, rather than necessarily elevate the nonhuman to a human status, such a symmetry can easily render humans to be little more than a behavioralist reading of their actions.[10] In this vein, there is also a tendency for ANT's human actors to be rendered mute. Actors instead come to be known principally through their manipulations of machines and other entities and in language performances only to the extent that they index the immediate situation. Language production and speech acts in particular have long had an ambiguous and underdefined role in ANT, something I critique in chapter 4 in terms of the materiality of rumor. Matters of human motive and meaning, which are so often made apparent linguistically along with matters of enduring human history are what tend to fall to the wayside in the prioritization and circumscription of this form of materialist accounting. These particular silences and omissions as a tendency of such an analytical framework are a source of concern in light of the book's task of chronicling a population already prone toward invisibility. It is something I attempt to compensate for in the analytical framework proposed in the next section.

In the end this book is not strictly an ANT study. To address the alternate agenda I bring to the current case, I do something very un-ANT-like by carrying forward a regionally distinctive history that is fixed in the African studies literature revisiting this history with contemporary observations from the field. I raid the stores of the contemporary ethnography of Africa to enroll a conceptually rich body of work on orality and rumor, commentary on capitalist consumer culture from the periphery, colonialism and its aftermath, cosmopolitanism and youth culture, and the spread of world religions.

Reconceiving Users in Global Technology Studies

In this section I outline the analytical framework that structures subsequent chapters and offer it as a possible guide for scholarship in technology studies moving beyond the Euro-American settings of the field's early work and into the world's diverse peripheries. This book offers a contribution to the way the user is conceptualized in STS by considering this special class of invisible users. Certain adjustments to the prevailing analytical models of the user are necessary to render such a population more fully visible within the narrative of this book as well as to STS theory. In particular, this entails loosening up on the tendency to account for users in ways that are too narrowly circumscribed around their direct engagement at the human-machine interface. In a structured way I propose an expansion into other relevant cultural formations and to the broader political economy of Ghana as a nation-state and of Ghanaians in a global order,[11] showing how this comes back around to shape the experience of users at the machine interface.

The relationally material or material-semiotic logic I have described and advocated for has some precedent as a framework for user studies. The key examples are Woolgar's technology-as-text analogy and Akrich and Latour's similar notion of the inscription and description of technical objects (Woolgar 1991; Akrich 1992; Akrich and Latour 1992). These models define the user principally in relation to two other entities: (1) the developers of the technology and (2) the material form of the technology itself. Developers are understood to write the technological form, inscribing in it their ideas about the user and his or her expected and preferred behavior. Users, for their part, are credited with some degree of interpretive freedom. They read the material artifact and, as these models suggest, either conform to or alternately resist the developer's inscriptions. An influential pair of studies has used this approach in African fieldsites. These studies highlight the flexibility or rigidity built into material forms and how this affects whether such technologies can be successfully appropriated. A photoelectric lighting system designed by a French NGO for villagers in an unspecified African country (a design depicted as rigidly designed and closed off to prevent repair or alteration by users) is given as a case of design failure (Akrich 1992). By contrast, the built-in flexibility or "fluidity" of the Zimbabwe Bush pump (for pumping well water) is the source of its enduring use through the decades according to De Laet and Mol (2000). Both cases deal with user populations for which the item was intentionally (if misguidedly in Akrich's case) designed. This triadic model of developer-technology-user and the questions it generates about user acceptance of technology and the capacity for user agency is an invaluable initial reference model but needs to be tinkered with and expanded to accommodate the case of youth in Accra's Internet cafés.

What hinders this triadic model's broader applicability is the political-economic context of design and use that it presumes. This model is framed around customer-oriented design and engineering work for known markets. By contrast, in the Internet cafés of Accra one finds a space of access and use shaped, in part, by gray market processes that divert a technology (imported secondhand computers) to a user population and setting of use wholly unanticipated by its developers in the US-centric high-tech industry.[12] Therefore, the first adjustment to the model that I propose is to deprivilege the developer's role and to excise this language of conformity or resistance, which becomes meaningless when users are so absolutely disconnected from design and development processes. Users in Accra, for their part, contend with newly available material capacities of the networked computer in whatever way is flexibly accommodated. To speak of this as resistance implies a tie to production processes and an ability to intuit the mind-set of the developer, which is particularly inapplicable in this case.

With the developer's role thus set aside, what I incorporate into the model instead is a broader set of proximate roles—fellow users at the café, Internet café operators, and further afield, family as well as community authorities such as elders, church pastors, or teachers. These figures all play their role in sharing interpretations and collectively materializing the technology. The neglect of the broader social world of users is an established critique of this triadic model and the notion of technology-as-text (see Oudshoorn and Pinch 2008). It furthermore reflects another part of the implicit political economy of the triadic model—an assumption of individual engagement with the technology in private spaces, an arrangement that follows generally from equipment ownership. In shared and public access settings, such as the Internet café, where two people are often seated together before the screen speaking with one another in person while they travel online, where a user may peek at other adjacent screens, and where a nosy or helpful Internet café operator might direct use, the narrow focus on the isolated user contending with the machine interface alone is more apparently unsuitable.

This brings us to another challenge of the technology-as-text analogy and of its broader applicability stemming from the burden it places on the technological artifact to be internally coherent and self-communicating.[13] Such an analogy situates direct material manipulation of the given technology as the principle and definitive mode of sense-making. Technology's material form as a kind of text or script suggests that it has an inherent temporal dimension, an implicit sequencing and linearity. Yet the initial experience of entering the Internet café, as recounted by young Ghanaians, drew attention instead to the confusing multiplicity and simultaneity of the computer screens' interface elements, software applications, and peripherals (keyboard, mouse, monitor, Webcam). Of these first experiences, youth often described how the initial work of parsing and prioritizing interface elements came through other sorts of scripts. These young Internet users generally mentioned the key role played by a friend, cousin, teacher, or café attendant who offered a verbal characterization of the technology; physically demonstrated its use; and indicated where to start, the sequence of steps to take, and perhaps most important, a sense of why one would bother to use it at all. The script of the technology as communicated in its material form (to the extent that there was one) appeared to be crucially propped up by various other scripts that were in turn written by this great variety of proximate actors. A key consideration in the chapters that follow are the diverse scripts of this sociocultural setting with a focus on some key formats—namely rumors and church sermons—that have not yet been seriously considered as elements in technological sense-making.

Incorporating these necessary adjustments to a prevailing material-semiotic model of user behavior, interpretation, and agency, what I propose as an alternate framework begins with but then spirals outward and away from the human-machine interface (see figure 1.1). A spiral is an apt image because its outward movement occurs simultaneously with a notion of traversing back and forth across the social world under study. It implies

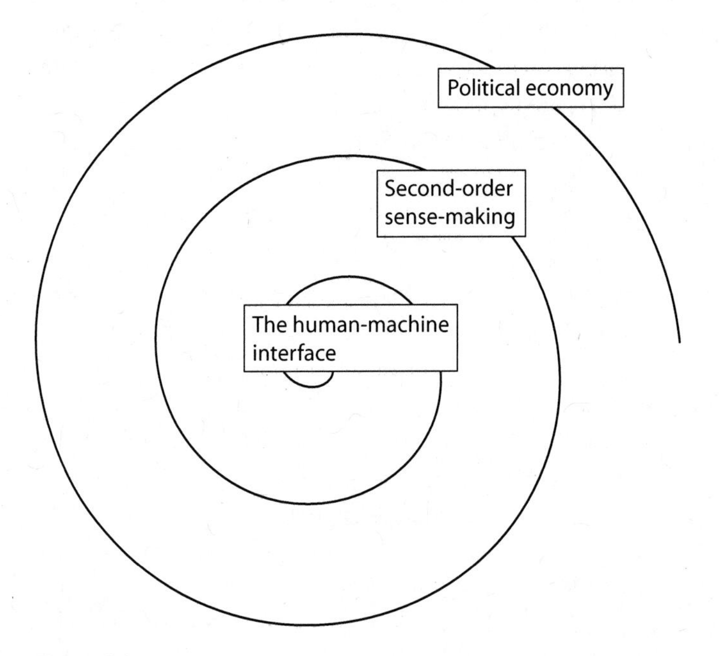

Figure 1.1

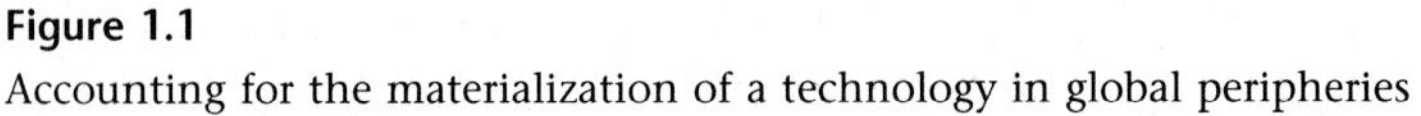
Accounting for the materialization of a technology in global peripheries

neither a hierarchy nor the containment of one level of analysis within another. The initial level of analysis, the center of the spiral, begins in proximity to the interface attending to human-machine engagements. In this particular case, attending to these engagements means also considering how computers, as a portal to the Internet, are situated within the space of the Internet café and how the Internet café is itself situated within its urban surrounds as considered in chapter 2. Additionally, what young Ghanaians do seated before the screen, how they present themselves to others they encounter online, is examined in chapter 3. These offer a foundational understanding of what the Internet has become in urban Ghana, its meaning as tied to its materiality in these particular spaces of access and use.

Moving beyond the immediacy of these Internet-use practices, in the pursuit of a broader cultural account of the technology, a second level of analysis is carried out in spaces of second-order sense-making. The practices carried out in these spaces deal with the technology in question but without direct manipulation of technological interfaces. The possible sites may cover the breadth of a society's cultural formations including speech genres, popular culture, educational and religious institutions, and forms of governing. Dwelling in these spaces illuminates additional cultural trajectories that implicate the Internet. Here is where one finds many of the supporting scripts that contribute to the coherence of the machine interface. This work away from the interface does not mark a departure from a materialist form of accounting but involves extending it into new arenas. In particular, maintaining an agnosticism about the relative durability of bodies (the somatic) and language performances (particularly speech acts) in relation to the built environment and technical artifacts helps to unsettle the assumption that the computer interface will always and automatically be the primary site of sense-making and material manipulation of the Internet. The primacy of the interface implied by the triadic model is instead treated as a question for empirical consideration.

Moving still further outward, the third level in this framework considers the distribution, refurbishment, circulation, and disposal of the technology in question. Delving into the domain of exchange complements the focus on the consumption of the technology via its direct use at the human-machine interface and in practices of second-order sense-making. Clearly such use is possible only when arrangements are in place to facilitate access. And how access is thought about, problematized, planned, and accomplished in urban Ghana shapes the specifics of how the technology is interpreted and used. Accra's Internet cafés were equipped with second-hand computers previously used in schools, businesses, and homes in the United States and Europe. A key role was played by Ghanaian transnational family businesses, which through a process of ad hoc and opportunistic trade constructed and channeled a flow of surplus technology from economic centers where the relentless course of technological upgrades and advancements deemed them obsolete and waste. The small, neighborhood Internet cafés throughout Accra were also the effort of small-scale entrepreneurship by Ghanaians locally and in the diaspora. Neither high-tech corporations through marketing or advertising campaigns nor the instructive and educational direction of planned public access campaigns of the state or by foreign aid agencies entered into this process. One result was a certain degree of interpretive freedom among young Ghanaians who were left to make sense of the technology without a human or institutional mastermind overseeing, guiding, or policing their efforts. This was one crucial way that the political economy of Internet access provisioning was consequential to the unfolding of direct human-machine engagements.

A conscious effort to work through these levels constitutes an approach to material-semiotic analysis that is more flexibly relocatable and adaptable to the full diversity of political economic processes around the world that bring technological artifacts together with user populations. As a demonstration of this framework, the chapters of this book are organized as a movement along this spiral and through a series of sites starting from the center. Each contributes to an understanding of the social world of youth and the processes that distinctively materialize the Internet in the Internet cafés of Accra, Ghana. These sites include the built and configured environment of the Internet café within urban space (chapter 2); virtual, online spaces (chapter 3); the social imaginary generated through rumor (chapter 4); churches and other spaces of religious or spiritual practice (chapter 5); the cloistered sphere of development work (chapter 6); and finally marketplaces (and dump sites) where computing equipment circulates before and after its deployment in the Internet cafés (chapter 7). The point is to define with progressively greater scope and from a variety of angles the transecting local and global processes that materialize the Internet in a particular way in Accra.

The first two chapters are the most straightforward as material-semiotic accounts. They illustrate how attending to interfaces—the meeting of spaces with material resources and human populations—can yield a better understanding of the surprising extensions and creative redirection that users undertake as well as the hard reality of their position in the margins. Chapter 2 introduces the populations and fieldsites that are the subject of this ethnography. It considers the way the Internet café is distinctively situated in the urban space of Accra. The Internet café as a space apart from the urban fray is accomplished through the maintenance of its threshold and the ambience and décor of the interior space but also through the kinds of interactions the equipment lends itself to. The materiality of the Internet café as a grounding for the social pressures and aspirations of urban youth produced a space of traveling through rather than of face-to-face sociability. With these observations I call into question the notion that Africanization or localization is the outcome of the successful appropriation of a technology of foreign origins. Youth who frequented the Internet cafés in Accra preserved the foreignness of the technology, resisted rendering it mundane, and in doing so embraced the suggestive possibilities of the Internet as an indeterminate entity.

Chapter 3 looks at the push and pull between young Ghanaians and the foreigners they encounter online and how the Internet as a mediation technology intervenes in this process. In particular, this chapter considers the innovation of Internet scamming known colloquially in the West Africa region as *419* or *sakawa*. Scamming activities are contextualized and explained in this chapter in relation to broader patterns of interactional breakdown in online cross-cultural encounters between young Ghanaians and their foreign chat partners. Text-based chat conversations or emails, though narrowly channeled through a low bandwidth and a time-constrained connection, were nonetheless a setting of more immediate and direct intimacy with foreigners than was regularly possible for young Ghanaians before the arrival of the Internet. However, this chapter also shows how the particular materialization of the Internet historically and the particular mediation of interactions between Ghanaians and foreigners worked against the creation of tolerant and mutual relationships that these youth so often keenly sought. This chapter handles the way young Ghanaians experienced a kind of social and interactional marginalization online and their distinctive response to it.

The next two chapters delve into what I have termed *second-order sense-making*. They consider specifically the critical question of how an interest in the technology was initiated as well as continually maintained and motivated from a socioaffective perspective. Chapter 4 examines the way anxiety and moral uncertainty around the newly arrived technology in Accra came to be handled through storytelling. In this chapter various rumors about Internet scamming that were in widespread circulation among youth who inhabited Accra's Internet cafés are considered as a source for local understanding and conviction in how the Internet works. In so doing, the analysis challenges a notion of speech as weakly ephemeral in contrast to the hard consequentiality of things, a characterization often implicit in the way theories of materiality in STS are applied to cases. Rather, the circulation of rumors, the way they compel retelling, indicates how different speech forms exhibit different degrees of durability. In particular, the narratives unfolding in rumors about "big gains" young Ghanaians had made through the Internet were important to maintaining the equilibrium of Internet use in light of users who often had disappointing experiences and faced indefinite delays in realizing material gains from these practices. Stories that reconciled moral questions (for example, by placing ultrarich figures conceived of as impervious to harm such as Oprah Winfrey and Bill Gates in the position of scam victim) highlight how the question of understanding the Internet was not simply about effective use but also about creating a morally sound portrait of one's role within it.

Chapter 5 describes how the functioning of the Internet and its apparent breakdowns came to be distinctively understood in a society in which supernatural forces are widely believed to operate in the everyday world from the "realm of the spirit" as one preacher put it. This chapter is principally about efficacy rather than morality in keeping with the main bent of a metaphysics promoted especially by the Pentecostal Charismatic Christian sects that are currently flourishing in Accra., From around 1979 such churches began to grow rapidly in urban Ghana becoming an increasingly visible urban presence through the 1990s. They promoted what has come to be known as the prosperity gospel, that one's properly aligned faith is rewarded materially and financially in this life rather than in the hereafter (Gifford 2004). In recent years, this has come to be related directly to the success of young Ghanaians' Internet-based social networking through emerging notions of the way such forces traverse electronic links.

The failure of the Internet to function as desired, including breakdowns in online relationships, might be attributed to such forces and resolved through rituals of the church or even visits to a fetish priest or mallam. The argument of this chapter refutes an aspect of digital age rhetoric, specifically an excessively sociological and disenchanted notion of how religion operates in the digital realm which reduces religion to no more than a form of affiliation and a community of belonging.

Finally, chapters 6 and 7 move further outward to consider some aspects of the political economy of the Internet cafés in Accra. Development institutions and aid agencies are very much a visible presence and produce an influential and ubiquitous discourse in this part of the world. In recent years agencies such as the UN, United States Agency for International Development (USAID), and others have taken a role in promoting the Internet and other information and communication technologies (ICTs) as tools of development. Chapter 6 explores the rhetoric emerging from development institutions that imagine what the Internet and other new digital and networked technologies will inevitably mean to marginalized populations in the Global South. In this chapter, I recount my participation in a relevant UN event, the World Summit on the Information Society (WSIS) Africa regional conference that was held in Accra in February 2005 in the midst of my initial period of extended fieldwork. This chapter considers the particular appeal of the information society as a unifying concept around which to organize a world summit. The rhetoric at this WSIS event was mapped out in the performances of participants and the documents that came out of the proceedings. Altogether they portrayed a dematerialized image of unencumbered, global information circulations, an image absent any sense of costs or trade-offs to joining the digital age. The efforts of organizers and participants at the WSIS demonstrated a profound disconnect between the depoliticized and impersonal framing around *information* and the *information society* of this institutional expert culture versus the perceptions of an active, racialized, and politicized exclusion online that was the major concern of Internet café users in Accra.

Chapter 7 functions as a kind of epilogue to chapter 6 by considering the hard materiality of computing technology and the Internet as it became increasingly apparent to ordinary citizens and the political elite in Ghana. The overtaxed electricity infrastructure, the influx of computers and other electronics as a burden on waste-handling systems, and the financial flows necessary for the business of enabling connectivity were part of an emerging awareness of the true costs of gaining Internet access in Ghana. This chapter considers import policies and practices as shaped by national interests as well as by external agents, such as the US-based NGOs that have recently begun to push for reform in the export of used computing equipment to places such as Ghana. These goods have been recast as toxic waste through the activist work of NGOs bolstered by Western media coverage. This chapter considers the perspective of local entrepreneurs, specifically Ghanaian computer importers and scrap metal dealers, and how they navigate and enroll resources within this structure of opportunity and regulation. This shifting political landscape has potential consequences for the ongoing viability of Internet cafés as small businesses in Accra.

Finally, the account comes back around in the conclusion to relate this case of the Internet cafés of Accra to some of the broader claims of digital age thinkers. During the course of the six years of this study, the Internet café scene in Accra changed along with the increasing visibility of Ghanaians online. The conclusion looks specifically at how geographically localized security threats (such as the scamming strategies associated with the West Africa region) are coming to be managed in network nodes located in the United States and elsewhere. I show how over the six-year period from my initial fieldwork in 2004 to my most recent trip in June to July 2010 a process of encoded exclusion has taken hold as practices deemed illegitimate (reflecting the peripheral position of this user population) led to country- or region-level blocking. In light of this new visibility we may ask whether the activities of youth in these Internet cafés and their pursuit of cosmopolitan longings and foreign contacts will remain viable. This chapter also delves into the recent debate over network neutrality in the United States, which has once again enlivened discussion of the founding ideals of the Internet as they relate to more recent trends toward the Internet's commercialization. This discussion touches on particular conceptions of discrimination and user and consumer rights that are propelled forward with concern principally for US industries, US consumers, and US federal regulation. There are possible, perhaps inadvertent, global consequences of such debates.

A prime challenge of studying any very new technology in its earlier stages of diffusion and uptake is of separating the consequential and enduring new capacities from the more faddish and transitory enthusiasms.

There is a well-documented history of the tendency for untamed rhetoric to erupt (ranging from optimistic fantasies to dystopian nightmares) whenever new technologies arrive that capture the public imagination (Nye 1996; Marvin 1988; Adas 1989). The grounding in lived experience offered by ethnography is a vital tool for addressing this challenge. Underlying the more specific study of the Internet, youth culture, and urban Ghana undertaken in this book is a broader concern with understanding the possibilities and implications of materially mediated human encounter, an essential matter of interpersonal communication that is foundational to theories of society. In recent years a range of new possibilities for mediation have emerged, reconfiguring and extending human interchange in novel ways. Although the Internet is a focus of this book it is done with an eye on the larger conclusions to be drawn about connection and culture and also the enduring realities of marginalization in the ongoing global reordering of the twenty-first century.

2 Youth and the Indeterminate Space of the Internet Café

Immobility in a Mobile Age

This ethnography begins in earnest in this chapter with an introduction to the people and places that comprise the field. A small number of Internet cafés served as the starting point for my observations but this quickly expanded into various other sites and the broader urban environment the cafés were situated within. The regular inhabitants of these Internet cafés were for the most part male youth (ages sixteen or so to about thirty). Speaking with Internet café operators and owners, people in the neighborhood, and the families of some of these youth filled out the picture of the Internet café scene. The purpose of this chapter is to establish not only that certain Ghanaian youth desired and used the Internet, but also to begin to consider what specific capabilities of the technology appealed to them in the context of their broader life worlds. We may ask, if the Internet was the answer, what was the question?[1] Furthermore, what made the Internet café as a particular kind of space appealing in the context of the dilemmas and desires of the urban youth who spent their time and money there?

Among the young inhabitants of Accra's Internet cafés there was a widely shared fixation on making foreign connections and specifically on possibilities for travel overseas. International mobility, as they recognized, was a privilege very unevenly distributed among the world's populations. I was reminded of this again and again when conversations with youth in Accra turned toward the aspiration to travel, denied applications for travel visas to the United States or Europe, and money lost to "connection men" who claimed to have back-door contacts at the embassies. The reality of such enduring immobility in this part of the world seems to exist in the shadow of the more widely remarked-on global trend toward deterritorialization, "the loosening of bonds between people, wealth and territories."[2] This is often depicted by scholars as an uncontrollable process resisted or coped with through reformulations of locality (Appadurai 1996) or identity (Castells 1997). Yet, the other side of this equation, that of the selective blockages enforced by repressive states or imposed by foreign governments, remains little considered. Although the global view may highlight an overall acceleration, from a situated perspective this "loosening" is instead often experienced as dramatically asymmetrical. It may register first and most powerfully in certain domains (often media or commodity flows) whereas others remain stunted (especially labor flows). For young Ghanaians, exposed to an increasingly varied and global sense of possibilities (courtesy of the radio, television, and other mass media), their inability to reach those possibilities became all the more intolerable. Such an asymmetry threw into high relief their state of stagnation and exclusion. Following from this, the activities of youth in the Internet café and beyond were often oriented toward freeing up and routing around these barriers and blockages.

The consequent sense of marginalization felt among young Ghanaians in this state of "involuntary immobility" (Carling 2002, 5; Lubkemann 2008, 454) was registered spatially. It can thus be distinguished from the temporal notion of *abjection* that Ferguson proposes (with reference to post-independence Zambia) as a circumstance whereby access, privilege, and promise has been withdrawn, leaving behind only lingering, agonizing memories of what once was (Ferguson 1999). Youth in Ghana, too young to have experienced firsthand Ghana's decline after independence in 1957 to the famine and state bankruptcy of the early 1980s (Brydon and Legge 1996) are instead exposed to alternative lifestyles of distant and inaccessible sites mediated by mass media and return migrants. Yet, whether temporal or spatial, the agony remains.

Brad Weiss, in his examination of how fantasy is woven into everyday spaces by youth in urban Tanzania, refers to a "widening breach between the actual and the possible" (Weiss 2002, 100). I would suggest additionally a similarly widening breach between the *possible* (or perhaps *probable*) and the *imaginable*. Weiss links such a gap to evaporating formal employment opportunities in the wake of structural adjustment programs and the growth of an increasingly overcrowded informal sector. Although youth in

Accra similarly perceived a narrowing of options, they often related this to increasingly restrictive migration policies in the wake of 9/11. Muslim youth in Accra in particular perceived a growing suspicion and exclusion of their kind in the West. Altogether these observations call for a much clearer distinction between imagination and action. Under what circumstances do the two intersect and what are the implications when they do not? Appadurai argues that imagination is a "staging ground for action" (Appadurai 1996, 7). Indeed an imagination sparked and richly supplied by media imagery and other sources can motivate as well as illuminate pathways of action that were there but not visible before. Yet to imagine migrating abroad is not the same as actually doing so, as Ghanaian youth were keenly aware.

These observations about the experience of immobility in an age marked by movement sets the stage for the argument that follows challenging the expectation or inevitability of the Internet café as a space where the Internet comes to be localized or "Africanized" by youth. The Internet is relationally defined by the people who inhabit it, the furniture and peripherals (webcams, etc.), and the décor and ambiance. My claim is that the absence of an apparent effort to reduce the space fully to a set of prior meanings, to reterritorialize the Internet through the space of the café, should not be seen as failed appropriation or simplistically equated to enduring domination or Westernization via technology. Rather the signification of "foreignness" in the space of the Internet café reflected a tolerance among youth for the indeterminate. It was a particular strategy of handling the distance between imagination and action. It was furthermore an indication of an emerging social formation spilling over (in certain unruly and not yet fully processed ways) beyond the bounds of existing cultural schema.

Rather than subsuming the technology into an a priori social form, the methodological approach elaborated in what follows works backward from the site of technological engagement to trace out the most immediate social formations, those that implicate the technology as a generative source. The Internet café as a space must be considered in the context of the life worlds of youth in Accra and the broader urban terrain that such cafés are situated within. This chapter considers dispositions of inhabiting urban space as a lens for understanding this life world. How does the Internet café relate to the home, school, church, or streets in terms of how youth orient themselves to the setting or are made to occupy the space in particular ways? Many of the teenagers and twenty-somethings who inhabited Accra's Internet cafés had not yet completed schooling, had not established themselves in a line of work, and had not yet married. The Internet café was a space of (among other activities) competitive and collaborative games centering on the acquisition of foreign goods, foreign media, and foreign contacts in tests of independence and expanding control. Through their activities, carried out inside and outside of the Internet café, young people worked to produce the Internet and its functionality for themselves and for others. In the space of the Internet café, users managed sensitive social calibrations between one another as well as the technological devices they manipulated. Relationships within the space structured the technology's appropriation as it was negotiated by Internet café users, attendants, and owners. Users and nonusers looked to others for information, direction, encouragement, as well as permission to relate to the Internet and to inhabit the Internet café in particular ways.

On Method and the Internet Café as a Space of Traveling Through

It seems appropriate in a chapter tasked with a certain amount of scene-setting to recount some of the contemporary challenges of this urban ethnography and explain briefly my process for patching together a fieldsite to get at the phenomenon of Internet café use and the circumstances of urban youth. The heterogeneity, vastness, and complexity of the city of Accra itself was one of the challenges. The anonymity and confusion of such urban settings offering no "knowable social whole" makes it difficult for a fieldworker to reach that state of familiarity and a sense of comprehensive understanding that those in smaller, more contained communities eventually attain (Ferguson 1999, 18). This was further compounded by the ambiguity of studying the Internet, something users experience through a powerful sense of inhabiting and of spatiality but of a non-Cartesian kind. I have argued elsewhere for the notion of "the fieldsite as a network" as a contemporary strategy for managing such unbounded, ambiguous and heterogeneous terrain (Burrell 2009, 181). The Internet café as a space did not fully contain the social phenomena under study but rather served as an entry point (Green 1999; Couldry 2003) well-suited to understanding a certain population of nonelite youth in Accra. The space of the Internet café, with its frequent circulation of users and digital objects, spun broad webs across urban as well as virtual terrain. By following these threads in the style of multi-sited ethnography (Marcus 1998) I was able to maintain a concentrated focus on the emerging themes of youth, aspiration, and technological engagement.

In my first few visits to Internet cafés in Accra, I was initially struck by the very limited amount of proximate, face-to-face sociability that took place there. This was quite at odds with existing accounts of Internet café use in non-African geographies (Miller and Slater 2000; Wakeford 1999, 2003; Laegran and Stewart 2003). It also diverged with studies of public spaces of media consumption in an African context such as Weiss's observations of barber shops in urban Tanzania (Weiss 2002) or Larkin's consideration of cinema theaters in northern Nigeria (Larkin 2008). When they were asked, Internet users in Accra commonly stated that they had never met anyone in the café itself. Although I witnessed youth traveling through the cafés in groups, occupying it for a time with their noise and numbers, these were inevitably pre-existing groups. The key dynamic was one of *traveling through*. The Internet café was a departure point, either to online spaces or to subsequent physical destinations. It did not seem destined to become the sort of café-as-third-space thought to play a role in civil society formation (Oldenburg 1999). The business model and expenses of public access to computing as well as the interior arrangements of the café contributed to this. Given the expense of the Internet café, where users paid per minute or per hour to use the machines, young people felt a pressure to use the Internet with focus and purpose and, while on the clock, to block out and ignore any surrounding activity in the café. The money to go to the Internet café was often hard to come by and was not enough to support the regular patterns of visitation to the café likely necessary to generate and sustain social ties there. Those still in school and living at home managed to pay for Internet use by skimming from the daily "chop money"[3] their parents gave them for food and transportation. The typical arrangement of machines in the Internet café lined up side by side along the walls usually left users antisocially positioned with their back to the center of the room. Furthermore, the minimalism of these small Internet cafés featuring only machines and chairs in a 1:1 ratio and no additional table or seating space offered limited support for extraneous social engagement and interaction.

Technical constraints in the arrangement of the human-machine interface to the Internet also worked at cross-purposes to sociability. These users undertook their text-based interactions in silence and the single mouse and screen size was oriented to support work by a single user. In some cases this was reinforced by management as it was at a café called Lambos where a sign was posted on the wall asserting "PLEASE ONE MAN ONE COMPUTER." Scholars have previously noted the peculiar detachment between physical and mental presence when online (Hine 2000), something T. L. Taylor (1999) describes as "plural existence." Media consumers often experience the loss of an active sense of self and of transport to elsewhere that follows from immersion in the narrative of a very engaging television show or movie. There is additionally a profoundly interactive component in the spaces of the Internet, a tangible, physical experience of steering the course of the unfolding narrative. The hand moving the mouse corresponds precisely and instantaneously to the icon on screen. The immersion of online spatial presence means that, in the moment of interaction, sociability online becomes largely incompatible with sociability offline.

The reality of this incompatibility became quite obvious as I struggled to work out the pragmatics of my research practice in these early visits to the Internet café. The online social experiences of users were frustratingly difficult to observe, materialized primarily as scrolling text in a small window. My initial plan to sit down and observe over the shoulder of users proved unworkable. My presence as foreigner and my unknown motives seemed to be especially disruptive. When I approached Internet users in the café, rather than continue their online encounters, they would instead take me along on a how-to tour of the Internet that seemed to bear no relationship to what they ordinarily did while online. In observing what young Ghanaians did when they went online I could intervene from only one side of this social engagement and thereby risked obliterating the interaction in a quest to understand it. Taking users out of the space of the Internet café for an interview proved the best way to gain entrée to their difficult-to-observe, online lives. Consequently, these conversations with Internet users reflecting on and making sense of their experiences are a significant component of this ethnography alongside my subsequent observations of their chat conversations, of the Internet café space, the streets, and of other relevant social spaces for youth such as church services and homes.

Outside of the boundaries of the café space, the challenge of this ethnography was to foreground the most concentrated expression of an unfolding social phenomenon from the cacophony of the urban terrain. This process involved, in part, retraining my visual perception (as through a process of language acquisition) to draw the relevant distinctions in urban topography. This was made still more difficult by the marginality of youth that relegated them to seize on waste and surplus opportunistically—to fit themselves innocuously into the interstices of the urban terrain. I drew from interviews in which young people described their day-to-day lives in relation to space. The indexicality of these conversations, offering pointers to space, helped guide me in seeing anew the urban terrain.

One area of Accra where I spent time in and around the Internet cafés was Mamobi, a slum or *zongo*[4] that was (at least initially and to my foreign eyes) notable for the indistinctness of its terrain, for the open movement across boundaries and thresholds. Due in no small part to the municipal authorities' neglect of the community there was a mingling of sewers, garbage, and walkways; cars, pedestrians, and livestock; and residential and commercial spaces. Vehicle traffic flowed like water, only limited by impassable physical impediments. The encroachment of dust into interiors was constant. Buildings painted with a date and the initials "A.M.A." (Accra Municipal Authority) labeled the numerous informal, unapproved structures and dwellings. This reinforced the sense of an adaptive and transgressive relationship to space and the resistance to efforts from the authorities at enforcing divisions and allotments. I visited families in Mamobi who were living in windowless structures that were initially designed as storage sheds for nearby market vendors. Large compound houses initially built to house a single extended family were divided down into smaller and smaller chamber-and-hall units for tenants. A Christian family I met lived among Muslims in a small unit within a compound house shadowed physically and aurally by the mosque next door and the five-times-daily call to prayers. In a society with limited access to credit, buildings were assembled brick by brick as funds allowed and Mamobi was dotted with such unfinished structures, often inhabited with or without the permission of the owners. Modern commodities such as television coexisted with deprivation in more basic forms, such as lack of convenient access to clean water and toilet facilities. The family home of a young Muslim man living under extremely crowded conditions with mostly unemployed family members had, in the middle of the yard, a huge satellite dish beaming news broadcasts from the Middle East, installed by an Imam who was staying at the house. Against this intermingling, the distinctiveness of the Internet café was apparent in the more rigorously maintained threshold—beginning with the routine of sweeping dust from the floors each morning and sustained by keeping the doors firmly closed to preserve the cool air of the air-conditioning unit. These practices created a distinct segregation of this space from the outside.

An inability to properly see subtly marked out territories and distinctions in the neighborhood meant that at first I missed entirely the apparent organization and visible markers placed by youth to mark their presence in the urban terrain. Most significantly, youth groups erected signboards or painted walls with their name and motto, a practice whose traces I suddenly found visible everywhere once I was informed of its existence (see figure 2.1). In Mamobi these groups were often referred to as *bases* and were generally all-male and organized around soccer competitions. Some pursued entrepreneurial ventures in addition to sports activities and simply hanging out. One base, for example, had constructed a pay-per-use public shower for this area that was struggling with infrastructure problems. In another, members were collectively learning to repair mobile phones. Not to put too wholesome a spin on the trend, bases also had a reputation for smoking weed, engaging in criminal activity such as mobile phone theft, and for harassing girls who walked past. A related organizational pattern were the self-identified youth groups that, in contrast to the bases, could be mixed gender and were often focused more centrally on community service activities.

Consider two contrasting examples. First, a group of young Muslim men calling themselves "Friends Rest Society." I had heard by word-of-mouth of their existence and so one day I went in search of their hangout. On a Friday, when the pace of activity slowed down for Friday prayers in this predominantly Muslim area, I found the group of young men under the cover of an unfinished garage that housed a tro-tro. They sat crowded along a long wooden bench and on a set of folding chairs eating and chatting. The one who talked to me most extensively about his Internet use spoke also of his admiration for Russia, for its former superpower status, its weapons and machinery, and the exoticism of snow and gymnastics (which

Figure 2.1
A base in Mamobi called "German City"

he had watched on TV during the last Olympics). He went by the nickname "Moscow." The group did not yet have a signboard but were in the process of acquiring and painting one to mark more permanently their existence. About half of the members visited the Internet cafés together to chat with foreigners, play games, and read news and such spaces had become an extension of their sociable activities. In contrast to the emphasis on recreation signified by the name "Friends Rest Society," the "Avert Youth Foundation," its NGO-like name reflecting qualities of goal orientation and self-discipline, was founded and led by a charismatic and very intelligent young man named Farouk. The group held formal weekly meetings that I regularly attended complete with agendas and "rules of order." These meetings were held alternately in a Koranic school or in the back parking lot of the offices of an NGO. Their already erected signboard, standing at approximately eight feet tall and colorfully painted stated their motto "patriotism, endurance and hope" and slogan "Avert . . . rebellion is now!" along with their web page URL and email address (see figure 2.2). The

Figure 2.2
The Avert Youth Foundation with their signboard

Internet was their resource for learning materials and Farouk maintained a web page for the group to promote its presence and make its needs heard by the outside world. In this way, the Internet came to be used not just within the boundaries of the cafés walls but also in the streets and youth club meetings and in school yards and dormitories as following examples will illustrate.

Youth in Urban Ghana

The self-organizing and claim-staking of urban youth in bases and other groups in Mamobi can be further situated by considering the broader circumstances of youth in their relationship to the older generation in Ghana. As in other urban regions of African nation-states, the population in Accra, Ghana, skews disproportionately young due to high birth and death rates and patterns of rural to urban migration. Nearly 60 percent of Ghanaians overall are age twenty-four or younger.[5] Yet although youth are powerful in number, in practice, control over the most established and lucrative resources (i.e., land, capital, and government jobs) is determined and maintained by the older generation. Gerontocratic principles remain at the core of the social order of family and clan. Young people are expected to be in the service of their elders, to defer to their authority, and are socialized to show restraint in the way they speak.[6] Young people who lack well-placed older patrons to facilitate their success, blocked from established paths to advancement, emerge as prime candidates for experimentation with anything new that comes into social circulation. This is illustrated by the great popularity among youth of the newer Pentecostal churches in Ghana (Gifford 2004), emerging forms of artistic expression such as hip-life music (Shipley 2009), and novel business schemes. These schemes have recently come to include international multilevel marketing organizations[7] as well as more blatant scams promoting work or educational opportunities. In sum, what was not yet claimed by the older generation frequently was seized on by the young. The Internet café was a prime example of this type of new and unclaimed space.

Youth who frequented Accra's Internet cafés employed two main narratives to define their social category and present circumstances. One recurring theme was of the indefinite postponement of their ascent to adult status. This desired status was defined among these urban youth as employment, marriage, and the ownership of property and modern goods such as cars and electronics. For young Ghanaian men, marriage required a demonstration of economic viability starting with gifts given to a girlfriend and culminating in a formal engagement ceremony when gifts of money, clothing and yards of fabric, domestic equipment (often a sewing machine), and other items are given by the groom's family to the bride's family. Isaac, for example, who worked as an Internet café operator, complained that his current job was taking him nowhere and he couldn't make enough money to get beyond base-level subsistence. To address this stagnation he was making plans to travel to China to, in his words, "escape the poverty cycle" although these plans never quite materialized. Descriptions of this state of stagnation reflect a sense shared by youth of a profound lack of agency and of advancement through life stages as something that was far from inevitable.

In a second common narrative, young people described the unexpected interruption of their education or uncertainty about their eventual ability to resume schooling. An unequivocal enthusiasm and hunger for education was almost universal among young people encountered in the course of this research. Yet it was also common to experience the sudden evaporation of any hope of further schooling with the death of a parent, sudden change in family fortune, or lackluster performance in an exam. Money problems forced many young people to leave their schooling and enter into the labor force. Kwadjo, a professional videographer and Internet enthusiast, describes this in his family's decline noting, "[my mother was] a single parent taking care of about three children. It wasn't easy for my mom and when I was growing up my mother didn't have a stable business . . . at a point of time and she went down to [break] quarry stones and that was the most difficult and the saddest aspect of my life. Life wasn't easy so because of that I didn't use the normal time . . . to complete education. Sometimes I have to help my mother by going to sell, hawk." The reality of this was observable in Accra where one could see children and teenagers thrust into petty trade hawking pens, gum, bagged water, and tchotchkes at intersections and traffic jams throughout the city moving from one car window to the next. Chant and Jones (2005) cite census statistics from Ghana indicating that 58 percent of primary school–aged children were enrolled in school but dropping off precipitously to 31 percent enrollment in secondary school. Schooling was far from a certainty in Ghana especially at the higher levels. Yet to be in school was a singular marker of a young person's reputable status. The stigmatized category of "school leavers" to whom crime and disorder were attributed in conversation and the popular media in Ghana illustrates the social opinion directed at those who had lost this status.

Looking at these two narratives, of indefinite postponement and of unexpected interruption, underlines the theme of "liminality" or of "disappearing transitional pathways" (Jua 2003, 13) that leave behind a gaping chasm between childhood and adulthood in urban Ghana. It is a state of being between, not one thing or the other but on a threshold without any clear bridge from one side to the other. Exiting prematurely from school did not make these young people into adults and what they recognized as the rewards of adult status seemed to be perpetually out of reach. By contrast, what runs through many characterizations of youth by scholars commenting on general trends in Africa (Abbink and Kessel 2005; Cruise O'Brien 1996) or in more specific country studies (Comaroff and Comaroff 1999) is a focus on spectacular and disturbing acts of violence perpetuated by this population. What can be understood from the study of child soldiers, anarchic youth mobs, and political thugs should be seen as quite distinct phenomena from the circumstances of the average urban youth living in more mundane times. The eruption of violence as it is characterized in many accounts of African youth seems to perpetuate the exact opposite conclusion of what youth in Accra wished for me to understand about their position, the ways their agency was circumscribed and constrained rather than being prone to pernicious overflow. There is a fundamental disagreement in Ghana between elders who see the young as uncontrolled and of youth who see themselves as unable to act. This generational difference is also rendered in the way young people depicted the nature and intent of their activities in the Internet café and how their elders perceived the Internet café and its draw for youth.

Livingstone's work on how parent and child relations are mediated through technology adoption practices in the United Kingdom is a useful point of contrast to the experience of Ghana's urban youth (Livingstone 2002). Livingstone documents a discourse of safety that drives certain media and technology consumption practices of youth as they are facilitated by parents. The private bedrooms of the young are elaborately outfitted with media and communication technologies to keep young people safe in the family home and under parental surveillance. Livingstone relates this to a growing perception of urban public spaces as dangerous reflecting a central preoccupation of the "risk society" (Beck 1992). The monitoring activities of the older generation also shape how young people adopt new communication technologies in urban Ghana but in quite the opposite way. The draw of the Internet café was as a space where youth could escape this parental surveillance. The attitudes of the older generation in Ghana toward youth diverge markedly from this concern with ensuring safety underlining instead matters of discipline. Restrictions on youth emerged through the efforts of authorities to cultivate disciplined young bodies and minds, which is underlined at the annual events for Independence Day celebrations. I observed one such event on March 6, 2005, at the grounds of a regional school. Competing teams from a variety of nearby schools performed in uniformed marching squads to win awards based on their precision and conformity. Supervision and enforcement of behavior was not only confined to home or school but also extended into the streets and other public gatherings. I once observed Kwadjo, a young man who had no children of his own, smack a young girl firmly on the leg while she was sitting on the lap of her mother because she had not responded politely to his question. The mother thanked him then and there for his efforts to contribute to the girl's behavioral training and socialization. Although these checks on behavior were limited by a certain unavoidable anonymity of urban living, it was socially acceptable and expected that older members of the community would monitor and regulate the behavior of young people whether or not they were kin. The Internet cafés in Accra functioned as one of the few spaces where young people could escape this constant regulation of their activities. For their part, some elders expressed alarm at the possibility that such spaces could draw the young outside of this continuous space of monitoring and discipline.

Peer Groups in the Internet Café

Young people's relationships with peers and elders, their aspirations, and available free time all contributed to the appeal and popularity of Internet cafés among this group. In terms of free time, many young people faced periods of boredom and inactivity after they completed a level of schooling. There were long gaps when students were not actively enrolled in school but were preparing for exams or waiting for exam results to come out. An exam was taken after junior secondary school typically[8] at age fifteen. Again around age eighteen many students took the Senior Secondary School Certificate Examination or university entrance exams. Students took these exams in the fall but didn't receive their results until the spring. Those who did not do well enough to advance had to re-sit exams and wait again for results. Many were eager to find a way to fill the extra time. Isaac, for example, was offered the position of driver's mate on a tro-tro bus[9] by his father while he was waiting for his exam results. He readily agreed to the job noting that "I was lonely at home doing nothing." Many young people took computer and software training courses while waiting for their results. Spending time at Internet cafés was another way to fill this void. As Fauzia, an unemployed twenty-three-year-old noted, "Sometimes instead of sitting in the house, making noise, it's better to go to the café, to go and chat." Going to the Internet café was not only an enjoyable way to pass the time, but many young Internet users also felt the skills and foreign contacts gained might prove beneficial in the future justifying the investment of time and money.

In light of the discipline expected of young people at school, home, and most public spaces, Internet cafés held an appeal as spaces where they could escape the surveillance of their elders. Internet cafés were spaces dominated by young people and there was limited, if any, supervision by elders. Internet café owners were typically absent most of the time and operators were often of similar age, acting more like peers than supervisors. Young people could also arrange to visit cafés when they knew older people would not be around. For example, Gabby noted that he and his friends from boarding school[10] "will leave the school around ten, midnight because by then the price is low and then older people will not be in the net. . . . They will tell us to stop watching this pornographic." Internet cafés therefore became spaces for forms of mischievous, youth-centered, and peer-oriented behavior that they knew would undoubtedly be disapproved of by authority figures. These activities included watching and emulating music videos, flirting with foreign pen pals online, trying to hack the computers to get free browsing time, playing online games, or finding ways to obtain free things such as pamphlets, bibles, or CDs of computer games. Activities were sometimes configured by users to fill the space of the Internet café. For example, on one occasion I walked into an Internet café in the neighborhood of La Paz where everyone was gathered around a computer monitor watching *Top Gun*. On another, I observed a young man with some friends studying a video by the hip-hop artist Usher and emulating his dance routines.

Mischievous and playful activities were frequently carried out by young people in groups, often schoolmates on their way home from school or off campus from their boarding schools. Some discussed their forays online in the classroom and schoolyard thereby subversively expanding this terrain of youthful independence into spaces where they were highly supervised. Printers and scanners made it possible to do so in multiple formats. Gabby noted about his early Internet experiences, "I followed my seniors to town and we went to the café and then you see, when we got there . . . the pornographic pictures, sites, we have them plenty. . . . We go, scan them, then we paste them in our dormitories." Through these actions youth found new ways to challenge authority, not simply by escaping their surveillance but also by surreptitiously transgressing the regulated spaces of the dormitory.

Additionally, Internet café use served as a way of building social cohesion among peers. Kwaku, Daniel, and three other friends from school visited the Internet café together regularly and ended up forming a club. They formalized the group by going online to the British Airways Web site and signing up for the frequent flyer program receiving membership cards by mail. The British Airways program was called the Executive Club and so they took this name for the group. Although they never planned to use the cards, they were pleased to note that they could charter flights and receive other benefits from their membership. The group also developed a level of technical savvy by searching for what they called *illegals,* which were ways of hacking into Internet café computers to get free online time. These tricks spread by word of mouth in the schoolyard and quickly became overused so the group was constantly on the hunt for new illegals. Beyond technical tricks Kwaku, Daniel, and their friends invented a competition in which they would order free things online to see who could collect the most. They gave the example of a CD of computer games. The goal was not the acquisition of a computer game but the accumulation of a commodity that had traveled from a distant locale.

A theme carrying through the activities of youth online and offline was a fascination with abroad and with whatever evoked travel and connectivity with foreign lands. Airline membership cards and CDs or books by mail personalized with a name and address provided this sense of global interconnectivity that was far more compelling to young people than the gambling games the CDs contained. The Internet provided opportunities for making faraway places very tangible and personal—marked with the most intimate of labels, one's name. Accumulating mail was a way of demonstrating power spanning great distances and whoever collected the most material was declared the victor. This thrill was also evident in the most popular of Internet activities among youth—collecting pen pals. This activity was often conducted with such brevity and divided attention by Internet users in Accra that it became clear that a moment of contact rather than the content of extended conversations had value in and of itself (Slater and Kwami 2005).

Drawing on some examples of an external adult perspective on the Internet café practices of youth brings the broader reactions they provoke into focus. In a radio interview, Estelle Akofio-Sowah, the managing director of BusyInternet (Accra's largest Internet café), in 2005 argued,

> I was thinking it's probably our youth that have also, you know, taken up this cybercrime and 419 and again that's a national issue. Why is it that our youth have nothing to do, that they are free to take up these illegal activities? Why aren't they being properly educated? You know, why aren't they being given opportunities to work, to get work experience and to get a decent job. You know, I don't believe that all our youth are criminals. I think everyone has a good streak in them somewhere and wants to do a good day's work, but are there those opportunities for them? And again it's a national issue and when young people have nothing to do and they're bored, they get up to naughty things. [11]

In her comment, it is the state of being unoccupied that is attributed with producing undesirable, antisocial behavior. It also greatly concerned some older members of society that young people could be using Internet cafés to look at porn. A publication titled "The Word of God on Sex and the Youth" was handed to me by its author Victor Olukoju, a Nigerian man, at an evangelism and career-training workshop I attended in Accra. The booklet contained a chapter on the Internet that warns about Internet pornography:

Thousands of youth worldwide are already addicted to this ungodly practice. . . . People keep ungodly appointments to go and chat with boyfriends and girlfriends on the Internet. They occupy their time with such chats and do not have enough time to sleep early so they come to Church anytime they want. What sort of world are we living in? People pay money to commit sin! . . . The youths now steal and borrow money to browse the net just to "enjoy" these nude pictures and other immoral acts or films. [12]

As a religious figure Olukoju lamented the way the Internet café kept young people from attending church. Bernice, a young woman who had been educated about the advantages and disadvantages of the computer and the Internet at school, seemed to adopt this view of authority, noting one problem was that the Internet "made one lazy" and lamenting how in "Ghana here, most of the [students] we crack our brains, but since the invention of the computer, children go, instead of them staying beside their books, they just go there and waste a whole lot of time. . . ." Older adults saw the draw of the Internet café for young people keeping them up late and making "ungodly appointments"[13] as challenging the dominion of school and church over young peoples' time.

Although the prevalence of these activities was undoubtedly exaggerated in Olukoju's and other similarly alarmist accounts, it was occasionally possible to witness individuals and groups looking at pornography in Accra's Internet cafés. The comments young people provide about these activities counter suggestions of addiction by describing it as a matter of fleeting curiosity. Sadia described how some of her classmates—male and female—went to the Internet café to watch what they referred to as *obonsam* cartoons. *Obonsam* means *satan* in Twi and the phrase was a coded and humorous reference to porn videos that could be used in the presence of unwitting authority figures. These schoolmates would discuss and tease each other about what they had seen when back in class out of the earshot of the teacher. Young people were certainly aware that watching porn was a transgressive activity. Sadia warned, "There's a saying that anything that the eye sees it enters the mind and what enters the mind wants to be practiced. They looking at pornographic pictures . . . and they will like to practice and at their age I don't think you could do that." Although seemingly carried out in "public," viewing sexual imagery in the Internet café was likely the most private space available to youth.

That their elders would likely consider these activities frivolous, distracting, or outright harmful quite clearly served youth as a way of claiming personal territory, flouting social norms, exploring alternatives, and distancing themselves from the expectations of parents, teachers, and older relatives. These mischievous activities depended on such alarm and disapproval among elders. This reaction was the sign of having escaped the hold such authorities had on young peoples' minds and behavior. By asserting an awareness that this activity was considered harmful, they marked their defiance and agency. In this way pornography viewing and other transgressive activities in cafés served as a reaction to what Willis describes as "institutional and ideological constructions of 'youth' which privilege certain readings and definitions of what young people should do, feel or be" (Willis 1990, 13). Internet cafés were spaces where young people were able to contravene these constructions. They managed to walk a fine line between pursuits that were innocuous or hidden (thus permitting their survival) and those that were visible and transgressive, thus asserting their separation from the authorities and institutions that attempted to contain them. This was evident in the signboards and the occupation by bases of interstitial spaces alongside streets and in unclaimed buildings. It was also apparent in the occupation of the new and unclaimed space of the Internet café. Spilling over from the café itself, such claims on territory expanded through the transport of small items including papers, printouts, stories, and sayings into spaces of greater constraint such as home and school.

The Deterritorialization of the Internet Café

Past studies of Internet cafés have often focused on the kind of social setting constructed intentionally or unintentionally through the arrangement of space (Wakeford 1999, 2003; Laegran and Stewart 2003). Wakeford described the Internet cafés she studied in London as "landscapes of translation" where technology is "re-territorialized" by staff and customers. In an Internet café that targeted a female clientele she observed the efforts made to demonstrate a compatibility between women and technology through décor, graphic design, and staffing practices (Wakeford 1999, 197). Internet cafés in Accra, however, were experienced in quite the opposite way. The interiors of many of the smaller Internet cafés in Accra had a homogeneous quality. Their spatial design was not elaborately scripted.

Users perceived these spaces as providing a suggestive dislocation from place but without a parallel process of re-specification. Internet cafés did not translate technology into a distinctively Ghanaian visual or verbal vernacular; instead they served as places of escape (Slater and Kwami 2005), a departure point transporting customers, often to someplace indistinct but a place that was definitively not located in Ghana. Walking into the Internet café, customers were suddenly transported from a hot, tropical climate to the air-conditioned environment required by these machines from the cool North. Décor (if it was present at all) often made use of foreign imagery. For example, one Internet café was inexplicably decorated with very large poster-sized pharmaceutical ads that the café attendant said a friend had brought back from Germany. In Mamobi, a café had images of Mecca and other important Islamic sites displayed on the wall. The naming of cafés also reflected this deterritorialization. Sky Harbour in the La Paz neighborhood was named for the airport in Phoenix, Arizona (but notably the spelling of *harbour* is according to British rather than American convention). Mobra International Ventures was named to reflect the owners' not-yet-realized aspiration to become a globe-trotting businessman. These names are indefinite and suggestive rather than pointing to particular, well-known sites. They generally and ambiguously evoked the nonlocal.

The specific activities carried out online in the space of the café produced intensive, unresolved juxtapositions. To give a snapshot of this, while visiting Mobra International Ventures in early November 2004 I documented in my fieldnotes a list of what was being digitally consumed at the various machines. Two people were playing music very loudly—one playing Lee Greenwood's "God Bless the U.S.A." and another played a country-style cover of the song "Flying without Wings," a song originally sung by the British boy band Westlife. One web user was visiting the Real Madrid soccer team Web site. Another was viewing an "Islam Online" e-card announcing a customary marriage. He then emailed a friend about the upcoming presidential election in the United States and afterward googled "PCM job openings in the Gulf." Through such practices, Internet cafés as places accomplished a negation or transcendence of a self-consciously and narrowly defined Ghanaian identity rather than successfully accommodating and reinterpreting it. Instead the Internet café provided a canvas for imaginatively exploring cosmopolitan yearnings.

To be cosmopolitan, in a general sense, is to shake off the shackles of the local and consequently to have a weakened loyalty to a place of origins or home. Hannerz defines this further as a general tolerance for diversity and a desire to inhabit and immerse oneself in other cultures (Hannerz 1990). His definition relies, however, on movement through a sequence of definite places, a privilege of the international traveler, though he nods briefly to the future possibility of cosmopolitanism-in-place via mass media exposure. A distinctively African cosmopolitanism has been alternately historicized and defined by the process of detachment alone and less by what is put in place of home. This is apparent in recent religious trends on the continent as in an analysis in Malawi of Pentecostals as cosmopolitans who admit no possibility of a worldly home (Englund 2004). Ferguson situates cosmopolitanism within Copperbelt labor migrant flows out of rural areas and the resulting tension between urban workers and their rural kin when workers reject their obligations to rural social networks that would deplete them financially (Ferguson 1992). He further observes the performance of such an orientation in urban street encounters (of spoken exchanges and displayed fashions) as "those stylistic modes that refuse or establish distance from those pressures [of 'cultural compliance']" (Ferguson 1999, 212). This orientation also proliferated in Accra as a matter of social behavior and bodily fashion but furthermore in the cultivation of spaces and their atmospheres. It was externalized in Internet café–naming conventions, décor practices, music selections, the management of the threshold, of dust and of interior temperature as well as in the absence of effort at cultivation and a certain blank quality to such spaces. This cosmopolitanism embraced the indeterminate pointing to the innumerable and not-yet-materialized possibilities.

One example appearing at first to be an exception to this tendency toward deterritorialization in Internet cafés was the interior design at BusyInternet, the largest Internet café in Accra with one hundred public terminals. The product of local and foreign investment, the café was a profit-making venture guided by a social mission to cultivate technology education and investment in Ghana and throughout Africa. To demonstrate compatibility between African ethnicities and cultures[14] and technology was part of this mission and aided by financial resources that far surpassed what smaller cafés were capable of. Management made efforts to decorate the space with imagery of Ghanaians using technology. Local artists also had their work displayed and sold in the space. The attached restaurant and café served as a very hip hangout with a club atmosphere at nights that was popular with young people providing social interaction beyond technology use.

BusyInternet was a space where compatibility between Ghanaians and a global technoculture was self-consciously cultivated and put on display. It was common to see Ghanaians of a variety of ages using some of the most up-to-date technology gadgets there. BusyInternet hosted its fair share of teenagers and young men in groups but over time the management had implemented several new regulations to create a space where a diverse clientele would feel comfortable and welcome. This often came at the expense of a younger, rowdier clientele. The Nima Boys, presumed to be from the rough neighborhood of Nima, gained notoriety for rowdy behavior and fraudulent activities. They were gradually discouraged from using BusyInternet as their constant hangout. Management at BusyInternet eventually closed off access to all secure Web sites in an effort to prevent credit card fraud from taking place in the café. Rowdy youngsters did not disappear from the space but were forced to keep a lower profile. Furthermore, customers were allowed to sit only two to a screen, a regulation that was enforced, preventing groups from crowding around a single terminal. Undermining the space-dominating activities of young men seemed to have the effect that was desired by BusyInternet management of broadening the appeal of the space to diverse members of society.

Yet, the efforts at BusyInternet to recast the social and symbolic meaning of the space were not always interpreted by Internet users as intended. Benjamin, a university student who worked as an "Internet advisor" at BusyInternet, described how the furniture at the café made an impression on him, "I never saw these kind of seats in Africa. I've never even seen. The only time I saw that kind of seat was in a magazine. My sister is a marketing personnel and they buy these kind of things for companies. So I've only seen it in a magazine, but I didn't know. I mean, a whole company can buy this much chairs, all imported, you know?" He defined the experience of entering the space as instant dislocation as he notes, "the first day I entered, I didn't believe it. I didn't believe it was Ghana." Although BusyInternet was ultimately quite successful in creating a space where a wider range of society felt comfortable and where the compatibility between Ghanaians and technology was demonstrated every day, to an extent the foreign quality of technology and its setting of use prevailed.

Conclusion

Youth in Accra's Internet cafés, in keeping with their aspirational outlook and a frustration over their involuntary immobility, formed an attachment to the Internet as a foreign commodity and as a mechanism for making contact with foreign lands. Though this fixation looks, on the surface, like a process of localization that has failed or remains unfinished, there is no apparent reason why the obliteration of foreignness in a commodity should be the assumed end point when such items move across conventionally defined cultural boundaries. The appropriation of the Internet by young users did not result in the tool becoming incorporated into established, everyday patterns of activity rendering it mundane. Rather than ameliorate the distance between the commodities' origins and their own by absorbing it into a local vernacular, this distance was kept on display, preserved in various ways as an aspect of the commodity. The escape and migratory fantasies of young people, the particular pressures they were under in relation to the older generation, and the concerns they had about their role in their own society as well as the ever-changing global order shaped this alternate process of appropriation.

The value placed by these youth on the foreign or non-African qualities of the Internet and by extension the suggestive possibilities of the indeterminate shows how successful appropriation of technology needs to be considered as a separate question altogether from its localization or Africanization. The analytical reflex that assumes adoption and adaptation of globally circulating commodities is a move toward strictly local, a priori, and determinate meanings suggests, for one thing, unwarranted assumptions about the homogeneity of receiving populations. The arrival of a new or foreign commodity often generates competing notions of how it is to be accommodated as distinct, colocated groups scramble to reconcile the threat and promise it presents. To offer one example, there is Larkin's study of cinema theaters in urban Kano, Nigeria, reclaimed as a space of Hausa moral ordering. The tension in this instance was between two groups—the entrepreneurs building the cinema theaters as well as the cinema-goers on one side and Hausa religious authorities on the other. A wish to constrain and contain interpretations and uses, to restore social hierarchies, was evident among the religious authorities through their attempts to regulate who could use the space and when (Larkin 2002). For cinemagoers, by contrast, the space was one of magical transport to elsewhere. Through watching Indian films (the most popular genre) viewers could "imagine alternatives to Western modernity and Hausa tradition" (Larkin 2002, 320). This particular example illustrates how the process of resolving new and foreign forms is not necessarily adequately framed as a push and pull between geographic polarities—Western and non-Western or north and south. Rather than an opposition of successful appropriation against alienation, instead we see projections of possibility and the indeterminate positioned against moral enforcement and the determinate. Yet both trajectories of appropriation of the same material forms are equally local in the sense of taking place by players within a proximate and intertwined social world.

Although this chapter has not precisely established how one might judge an appropriation of technology as successful, one way of deeming it a failure has been ruled out. Appropriation need not be simply the respecifying of form with regard to local, prior meanings but also includes these intentional efforts to leave such technological forms open to interpretation. It is perhaps the etymological conflation between alien and foreign that misguides. To be alienated is to be in a state of unresolved distance between self and the commodity.[15] In Accra, a restored sense of status and of being more a part of the world for youth who inhabited the Internet cafés came through the preservation of a kind of distance or foreignness in the space of Accra's Internet cafés through certain acts of doing something (decoration, threshold management) but also by doing nothing and thus leaving intact the suggestive indeterminacies.

The marginality of these users is apparent at multiple levels: as globally situated and locally shaped. Youth struggled with an immobility that was in part a product of the way certain foreign governments regarded Ghanaians (and other Africans) with a particular bias against these young nonelites (as liable visa overstayers). These youth also sought to define and claim spaces apart from the determination of local authorities, apart from parents, teachers, preachers, and politicians whose surveillance and disciplining were so all-encompassing elsewhere. This translated into a distinctive approach to inhabiting the Internet café and engaging online that focused on bypassing these blocks on agency, approximating a cosmopolitan self through electronic travels. These forms of marginality were an ingredient in the formation of contemporary urban youth culture. However, as chapter 3 will consider, this marginalization was not simply evaded by going online but was encountered in new forms in the cross-cultural encounters of Ghanaian youth in chat rooms and other online spaces.

3 Ghanaians Online and the Innovation of 419 Scams

This chapter travels into online virtual spaces to consider digitally mediated encounters between Ghanaian youth and the foreigners they met there. The practice of collecting and cultivating foreign contacts (in Yahoo! chat rooms, on dating sites, and elsewhere) was the most common and characteristic use of the Internet in Accra's Internet cafés.[1] The bond users began to develop with key chat partners or, alternately, the thrill of pursuing an ever-expanding roster of contacts compelled many youth to return again and again to these spaces. The cross-cultural nature of these online encounters offers some unique, new insights to the literature on digital communication and to theories of cyberculture that until recently have been limited by the rather narrow socioeconomic and geographic characteristics of Internet-using populations. To consider these practices also means confronting some of the consequences of material asymmetries (in Ghanaians' delayed arrival to the Internet and the expense and time pressure of their metered use) as well as the dilemmas of self-representation online among the globally diverse populations that now inhabit these spaces.

In this examination of the cross-cultural encounters of young Ghanaians online, I seek in particular to explain the seemingly unprovoked and inexplicable reactions and sudden exclusions they experienced in their pursuit of foreign contacts. Mastery over such a sociotechnical system for these youth turned out to be more than simply a matter of grasping technical features of the Internet and its interfaces. In addition, a working theory of *obruni* (foreigner or white man) and how he or she might be compelled to respond in desired ways was critical to certain practices of use. Such an effort inspired a range of mimetic techniques. In particular, the phenomenon of gender swapping online and other manipulations of embodied cues were part of strategies of persuasion in forms of Internet scamming that emerged in the region.

In the scholarship that considers online environments, the virtual has often been analyzed as a domain of less material interactions in comparison to the "real" world. Some have argued that online spaces of the Internet serve as a unique setting of interaction via pure, limitless symbolism (Hayles 1999; Poster 1995b). I join with other critics in finding the opposition between the symbolic and virtual on the one hand and the material and physical on the other wanting (Castells 1996; Robins and Webster 1999; Sundén 2003). These days, the online increasingly spills over into the more obviously material constitution of the (quote-unquote) real world.[2] Additionally, the emphasis on unencumbered, limitless self-constitution and interaction overlooks all the ways that the Internet's hardware infrastructure and standards, as well as Web browsers, chat clients, and other software, draws up a structure for users to act within that is no more or less escapable than the structure of, for example, the more obvious materiality of a bicycle.

To treat online spaces as material domains is not simply to highlight their limiting force. Such a perspective also gives greater consequence to the actions of all who occupy such spaces. The overall decentralized design of the Internet has meant that a significant role in the materialization of this technology has been delegated to users. Users do much of the work of filling the skeleton created by standards, software, and built infrastructure with content. In chat room conversations, email, and personal Web pages, Internet users in Accra leave behind enduring digital traces. The 419 Internet scam strategies associated with the West Africa region (as considered later in this chapter) create new material conditions of the Internet for potential scam victims in the West via the email lure. This substantial role played by users, their ability not only to read but also literally to write the technology, distinguishes the Internet as a media technology from previous considerations of media infrastructure in colonial or postcolonial contexts where it has been viewed as a unidirectional imposition (Larkin 2008; Headrick 1981). These activities are not limitless but neither are they simply ephemeral and inconsequential. To consider the Internet's constitution by users as a material process is to acknowledge these traces left behind; their existence beyond the user.

The aforementioned dilemmas of self-representation online among diverse populations of users relates to the tension between novel possibilities (for disembodied interaction in particular) and the tendency for users to seek out highly conventional signifiers (of sex, age, race, etc.) to manage the uncertainties of social interaction with strangers or to realize idealized identities (Slater 1998; Nakamura 2002). When the citizens of countries around the world go online, they do not effortlessly shed prior imaginings of people and places and convert to a new social category—the net denizen. The baggage of prejudices and preconceptions they bring with them into the online domain is another kind of structuring that endures in spite of the possibilities of limitless symbolic self-constitution online.

Chapter 2 depicts how the Internet was accommodated by Ghanaian youth without an attempt to self-consciously render it into a kind of Ghanaian vernacular. This absence of explicit localization, I argue, did not necessarily mark a failure to adequately appropriate the Internet. In this chapter I suggest instead that a failure to appropriate can be located in the way young Ghanaian Internet users struggled to generate digital selves that were simultaneously authentic and persuasive. Admittedly, the notion of authenticity has become something of an analytical quagmire. Assertions of what is and is not authentic are well worn in service of ahistoric or bounded conceptions of culture (as noted by Turner 1992; Prins 2002; Ebron 2002). The authentic is often reduced to what is of proximate origins or preserves or restores the continuity of a "traditional" culture. Alternatively, we may consider how authenticity is at issue in the way Internet users relate their perception of self against the self that they digitally represent for a foreign other. A vast, unbridged distance between these two selves is generally recognized by young Internet users themselves as inauthentic. This comes through in the comment by one Internet scammer about the reasoning underlying his strategy of purposeful misrepresentation online: "if I put my real picture or if I tell you this . . . my family background, my status, my financial background you [foreigners] will not even want to talk to me." A sense of resentment among Ghanaians followed from the burdensome caricatures of Africa and Africans imposed on them by the non-Africans they encountered online and that they perceived as consequential to their capacity for online enactment. Around the notion of authenticity postcolonial and media theory may be brought together in fruitful dialog to account for the capricious encounters experienced by young Ghanaians in their newly realized state of connectivity.

Breakdowns and Disillusionment in Online Cross-Cultural Encounters

A description of the look and feel of online synchronous chat (the main application supporting young Ghanaians online activities) will provide a sense of context for the unfolding online relationships pursued by these youth. Recent anthropological analysis has turned to some of the highly visual and immersive graphical settings of virtually embodied role-playing such as Second Life (Boellstorff 2010) or networked gaming environments such as World of Warcraft (Nardi 2010). However, bandwidth requirements make this sort of pursuit still beyond what is feasible in Accra's Internet cafés.

Chatting of the sort young Ghanaians participated in took place in the austere environment of a series of text windows. The client software they used varied. Yahoo! Messenger was dominant and MSN Messenger was another popular choice. Some more obscure sites such as Hi5 were also used and more recently Facebook has begun to gain users in Ghana. The essential components of such applications were reliably similar. First, there was a directory of existing contacts, visible on logging in. In Yahoo! Messenger this served as a kind of address book of people the user had befriended. The list reflected through color and icons that current status of the individuals, whether they were online or offline. Selecting one contact from the list would open a separate window. This window was split into two blank sections, top and bottom. At the bottom was a space for the users to type words, their own contribution to the chat conversation. The top section was where the unfolding conversation appeared with color coding to distinguish one speaker from the other. Emoticons (such as 😊😘🤔👍) were available from a menu and changes in font and color could be employed by users to break up the textual monotony.

This chat interface was no more richly featured, no more apparently immersive or obviously environment-like than word processing software. Supplementing text with very brief phone calls or using a Webcam added some degree of richness to these online encounters. Despite the austerity, the descriptions given by users of the relationships they formed through this medium were often deeply felt. Clearly the meaning that could be carried between conversants, even when it was little more than typed text, was substantial.

Such a material environment may seem the ultimate neutral ground, a setting for two people to converse equally and openly. However, the delayed introduction of the Internet in Ghana meant that Ghanaian users entered into online environments that had, to some extent, already been staked out and normalized before their arrival. The evolving etiquette and ethos of such online spaces was set within and in reaction to Euro-American social norms and codes. By contrast, in their online forays young Ghanaian Internet users drew on a cultural logic and set of persuasive strategies that were close at hand. This involved certain assumptions about gender relations, dating rituals, models of patron-clientalism, and related redistributive obligations—ways that people are known to get by or get ahead in urban Ghana. One mistake Ghanaian users made was to assume that such a local moral economy would be coherent in a more heterogeneous global context. In other words, they did not always accurately perceive the social distances involved in online interaction or determine how to adequately navigate such distances. They were also hampered by the expense and time-sensitivity of public Internet access in Ghana, where users paid per hour of use. This led many young Ghanaians to pursue a rushed intimacy that had the tendency to be read by foreigners as transparently disingenuous online solicitation.[3]

The term *pen pal* was frequently and consistently used by Internet users to refer to online relationships with foreigners. In practice this term was inclusive of a wide range of types of relationships including same-aged platonic friendships, romantic relationships, older adults to appeal to for advice, patrons offering financial support, and even business partnerships. The use of this term comes from the prior experiences many young Ghanaians had with conventional pen pal programs (set up between schools) involving the exchange of letters with foreign students via postal mail. The practice of pen pal seeking reflects how the Internet was initially made sense of through analogy, drawing on familiarity with prior formats of mediated encounter. This ready template brought to the foreground certain possibilities for engaging with the new technological system but in the process obscured other possibilities.

There was a divergence between two ways of approaching these online relationships. Some pursued blatant strategies of gain seeking, approaching the Internet as a sort of gamelike environment. The particular experience of online communication, the blandness of text-based interfaces, certainly made it easier to dull the sense that there were actual, living people involved in such strategies. Foreign contacts were reduced to broad types and treated as interchangeable pawns in a scheme to get ahead. Daniel, for example, boastfully and humorously noted, "I take pen pals just for exchange of items and actually I don't take my size, I take sugar mommies and sugar daddies because if you take your size they are unemployed. So if you ask them something which is very huge, they wouldn't get it for you. So you have to go in for the grownups like 50 years, 40 and above. So they are actually working and they have the money so they can buy whatever you ask them." By contrast, an earnest yearning for cultural exchange and mutual support was another approach to the seeking of pen pals. The opportunity to correct misperceptions about Africa, one chat partner at a time, held great appeal for certain Internet users in Accra. Benjamin, a university student, described his Nicaraguan chat partner as completely ignorant about Africa and noted, "she thought Africa was a jungle and we live on trees. [laughs] I mean it was funny." Seemingly unoffended, he noted of this online friendship, "I really enjoy telling her about Africa and sometimes she doesn't believe, but she's forced to believe because I am here and I'm telling her. So it was very, very interesting, very educative. I really enjoyed it." The reference to "living in trees" was a recurrent one in interviews and circulated as a kind of idiom or saying rather than necessarily a direct recounting of conversations with foreigners in online chats. As a saying, it neatly encapsulated an irony young Ghanaians came to believe about the foreign other—that of their poorly concealed sense of superiority, their habit of reducing African populations to base subhuman status, and, at the same time, their astounding ignorance. For some of these young Ghanaians, the opportunity to present themselves as Africans who are technologically literate, their mere presence in the online space was in and of itself an effective way to counter such presumptions.

Many online encounters pursued by young Ghanaians, however, never got to the point of challenging and dispelling misperceptions of Africa in dialog with foreign chat partners. Rather many reported immediate rejections, mystifying and seemingly unprovoked animosity on disclosing their location or their nationality. Felix, an Internet café operator originally from Togo but now living in Mamobi, voiced his outraged disillusionment in the wake of such online mistreatment. He noted, "[I say] 'I'm from Africa' they say, 'sorry, bye bye' . . . they don't like black people. . . . We Africans, we know that white people don't love us. . . . Sometimes they can start to insult you, if say you are from Africa, insult you, talk shit." Moving a step further in his interpretation, beyond the insensitivity, disinterest, or animosity of foreigners, he suggests a more willful conspiracy is at work in this behavior: "They want to keep us, [from knowing] something about the computer. That's why they want us to stop going to the café." Fauzia, a young unemployed woman living in Mamobi, recalled a similar experience: "Sometimes when you mention your name and you mention Ghana they just say 'fuck you.'" As a young woman, she was also pestered frequently to expose her breasts to men on Webcams or met with explicitly sexual or offensive language. Her overall perception was that a certain crudeness was omnipresent and unavoidable in such online spaces.

Some of the more subtle problems of mutual intelligibility often emerged only after a long series of interactions broke down inexplicably. Fauzia, for example, met Yassim, an Egyptian man, in a Yahoo! chat room and in an early exchange took down his phone number. This was simply for the purpose of making his phone ring (referred to as *flashing*) as a way to further a connection and establish realness beyond text chatting. She said he told her she was a "good girl," a "wonderful girl," and gave her the nickname Angel. They chatted extensively, confiding personal matters to one another. Finally, after a period of developing connection, she made a request of him:

F: So "can I ask him for a favor" and he said "oh, why not I should go ahead." And I said "ok, my phone is giving me problems and I will be very grateful if he could send me money to get a better phone or if he could send me a new phone." [He asked] "that what?" And I repeated it again—I didn't see him online again. He stopped chatting, he disappeared.

J: Really? Wow. So what do you think about that?

F: So, since then anytime I go to the café I always leave him an offline message. "Oh how are you doing my dear, I hope you are doing well, I wish you all the best." Almost every day when I go to the café I'll send an offline message, "oh my dear how are you doing, I hope you are doing well, take care, be a good boy and be cool and then always pray to God."

J: Were you surprised that he stopped talking to you?

F: Yeah I was surprised because I asked him, the only thing I said is oh, if he doesn't have, if he can't help it's not a problem. So he shouldn't worry about that. So he shouldn't say because of that, he couldn't chat with me.

A number of young Internet users recounted this type of experience, one of digital shunning. Such enforced disconnection and avoidance followed a seemingly minor interactional misstep. Here the matter of authenticity can be read in another way as a matter of the depth or realness of a relationship that Fauzia perceived was strong in this instance but that was undermined by Yassim's reaction. Although for her the developing bond with her chat partner could reasonably imply the next step toward the tangibility of gifting, this was apparently not as self-evident from Yassim's perspective. Such a total and permanent avoidance seemed to be sparked most often by requests from young Ghanaians for assistance in the form of money or gifts. To put this in context, small transfers of funds between friends is a regular feature of relationships among youth in Accra.[4] It is guided by a cultural logic of reciprocity, that to give to someone in need today means that one will be in a position to receive later. In dating or marital relationships the regular transfer of small sums known as *chop money* from husband to wife is a sign of love and devotion. And in relationships between powerful, older, or wealthier individuals and younger protégés, a patron-client model is often operative whereby an affluent individual feels it a duty to redistribute funds or other forms of support within his or her social network to those who are less well-placed.[5] These well-established role relationships in Ghana, however, proved to be far less durable in the virtual domain. The true social distance between online pals revealed itself to be a gaping chasm when there was no possibility to confront, explain, or clear the air with a former intimate who chose to evaporate into the digital ether.

A certain ethos that came to be embodied in the material form of the Internet can be traced back to the social origins of the technology. Castells, recounting the history of the Internet's development, notes a distinctive emphasis on total and limitless freedom of expression among early founders of the Internet, the hackers, communitarians, and entrepreneurs involved in its engineering and development. This freedom was very much focused, however, on the individual's unconstrained capacity to act. The pursuit of fully flourishing self-realization online took precedence over collective interests (Castells 2001). Sometimes signified as a frontier, the Internet became a space where imposed regulations of a legal nature or written into software code were actively eschewed in favor of self-regulation by users (Turner 2006; Rheingold 1993; Ludlow 1996). Such a stance is maintained to this day in the permissiveness and trusting nature of technical protocols of the Internet that have been exploited in various ways by spammers, scammers, and criminal hackers.

Yet out of the freedoms proffered by the Internet has also emerged certain forms of antisocial behavior termed *trolling* or *flaming* (Lea et al. 1992) whereby aggressive, personally insulting, and abusive comments are left behind in discussion groups or otherwise placed to disrupt online discussions. Safely anonymous, physically distanced, text-based modes of interaction online eliminate certain forms of enforced accountability that usually temper such behavior in face-to-face encounters. For example, social aggressors often do not witness the reaction to their insults and abuse in online environments. In this way the prerogative of the individual once again seems to prevail. Meeting fire with fire, an "ignore," "hide," or "block" function has recently come to be programmed into a huge range of network applications, from email programs such as Microsoft Outlook to Yahoo! Chat to social networking sites such as Facebook. Such features encode a way to inflict exactly the sort of banishment experienced by many young Ghanaians online. It allows a user to expel a single contact absolutely, permanently, and with surgical precision from one's personal online realm, a capability unavailable in proximate, real-world societies in which individuals who have a falling out often cannot escape ongoing physical co-presence and overlapping social networks.

In this continual arms race of digital tools to support still greater degrees of self-determination, many of the pressures to accommodate and work through misunderstandings in social encounters are diminished. This tendency belies a certain early rhetoric on *cyberspace*,[6] a term now nearly out of use but that once signified online spaces as a wholly separate and distinctive sphere of activity. For some participants and scholars alike, cyberspace was a more permissive and tolerant social environment, one less prone to boundary drawing and social separation. One can conversely argue that the very real geographic distance and the peculiar properties of this channel of connection between conversants offer fewer checks on impatience and intolerance in online populations. The previous examples of young Ghanaians facing ostracism, ignorance, and insults point to this unfortunate reality of online cross-cultural encounter.

The 419 Email Scam and Its Variants

As the previous examples indicate, young Ghanaian Internet users struggled to make sense of their online cross-cultural encounters, facing the challenge of mutual intelligibility. When this broke down, what they often experienced was a kind of digital shunning. Within these broader circumstances of access and use, digital representation and interactional dynamics are the context that Internet scam strategies emerge from in the West Africa region. Scamming strategies shape and are shaped by online, cross-cultural interactional dynamics. The formats of these scams have evolved extremely rapidly to keep ahead of network security responses and the growing savvy of those they target. This was the activity demonstrating the most learning and evolution between my first visit in 2004 and my most recent one in 2010.

A distinctive genre of scamming over email, referred to locally as *419* (the Nigerian police code for fraud) or *sakawa*[7] (a vernacular term from the Hausa language) has come to be associated with the West Africa region, including Ghana, though it was originally (and still most strongly) associated with Nigeria. In the classic 419 email the author claims to be a wealthy former member of the corrupt Nigerian government needing to quickly transfer money out of the country. The extraordinary circumstances are explained by the death (often by plane or car crash or murder by political opponents) of a family member or business contact. The email recipient is asked to provide assistance often by making their bank account available for the money transfer. As a reward, the recipient is promised a hefty percentage of the gain often in the amount of several million dollars. Victims are asked to pay upfront fees for bribes or other costs and this money disappears. This model of scamming, also referred to as *advance fee fraud,* can be traced back to sixteenth century Europe (Glickman 2005). In Nigeria, such scams emerged out of an era of military coups, government corruption, and economic decline. Schemes were carried out prior to the Internet using fax, telephone, or postal mail through the 1980s and 1990s. The scam scenarios described in these emails are of actual activities of illegal money laundering carried out by the Nigerian military governments (Smith 2007; Apter 1999). Therefore, the narratives in these emails are perhaps more plausible than they seem to skeptics.

Stylistically these emails have been marked by their formality of address and dramatic narrative arch. A segment from one such email illustrates the detailed and concrete references to people, places, and events, the narrative recounting, and the formal language dotted with typos:

Kindest Attention: Sir/Madam,

I am Gerry Ogodu, The Secetary [*sic*] General to the former Senate President Senator Pius Anyim, of the Federal Republic of Nigeria, Though this proposal may come as a surprise to you as we have not met in any way before.

I got your contact address through your country business Guide and feel you will serve as a reliable source to be used to achieve this aim, by trusting under your care the total sum of Fifteen Million, Five Humdred [*sic*] Thousand US dollars (US $15.5M).

This money I want to invest into any business of your choice in your country was acquired through the Construction Of the newly completed Abuja National Stadium which was used for the just concluded All African Games (COJA) Abuja 2003. During the time the contract was to be awarded to a foreign ! contractor, I used my position and office then as the Parmenent Secetary [*sic*] to the Senate President to over invoice the sum of USD15.5M from the main contract sum.[8]

The success of such 419 scams relies on mass mailings of thousands of email addresses harvested very unselectively from the Internet (known as spamming) but potential victims are meant to read the appeal as personalized. They have been singled out as "a reliable source," evaluated as worthy of being entrusted with the task. The tone is often one of intimate disclosure to a trusted confidante. The technical properties of the Internet (and of the email protocol in particular) enable this front by allowing the email sender to conceal the list of recipients using the BCC (blind carbon copy) function or by automating the sending of many thousands of individual emails using mass marketing software or services.

Variations on this email scam include tales about lottery winnings, church projects, or business deals, all falling roughly into the same category—though variously playing on the greed, sympathy, or lust of targets. The content of such emails was not limited to the Nigerian political scene and authors often claim to be from other African countries. The conflict over diamond mines in Sierra Leone and Liberia and the seizure of white-owned farm land in Zimbabwe are some of the plot points used in these emails. Although Nigeria continues to be commonly identified as the source of these emails, the content generally ties them to any of a number of sites on the African continent.

The sorts of scam strategies I observed in Accra were similar to the Nigerian approach that employed misrepresentation and persuasive storytelling though did not precisely follow the classic 419 script. Instead they used strategies based more often on bodily seduction for persuasive impact. Internet café–based scammers in Accra often initiated contact with potential victims through chat rooms or dating Web sites instead of by email. Rather than pose as a wealthy government official, young men typically took on an alternate female persona while online in order to lure a foreign boyfriend. This came to be referred to colloquially as a *come-and-marry* scam. In producing these identities, scammers recruited a variety of disparate elements including female friends and family, digital photos of black models, Web sites with "love quotes," fake ID cards, Webcams, and other resources. Invariably the primary scammer was male, although he might recruit a sister or a female friend to assist by taking phone calls from the foreign boyfriend and by sitting in front of a Webcam while the scammer typed. She thereby served as the face and voice of the scammer's female persona. These scammers typically invested significant time in building rapport before suggesting any financial dealings. Once the "boyfriend" was properly seduced, the scammer would invent a scenario. He might ask for money to pay for travel so that they could meet in person or he might claim a family member was gravely ill and ask for help with medical expenses.

The young scammers I spoke with in 2005, by their own admission, saw few if any gains from such strategies, though this had changed by 2010 as I consider briefly here and more extensively in the book's conclusion. The particular population of scammers considered in this chapter represents a lower rung of amateur, free-agent Internet scammers. Their approach should be distinguished from what evidence suggests is, in its most effective forms, a type of organized crime in which bosses direct underlings in the pursuit of victims strategically and efficiently and with significant technical savvy (Glickman 2005; Apter 1999).[9] By contrast, these freelance scammers had not yet embraced the possibility of sending out mass emails (spam) for finding susceptible victims and so struggled along with individual scam targets laboriously developing rapport and an illusory relationship before springing a monetary request on them. This practice was extremely time intensive. As the previous section showed, these sorts of relationships, painstakingly developed and maintained, may balance on very tenuous grounds of mutual intelligibility. The foreign targets' belief in the authenticity of the scammer's online self-presentation was critical. If this was violated the scammer might find himself suddenly disconnected and thus the investment of time and money at the café would be totally lost.

Disembodiment and Gender Swapping as a Scam Strategy

The practices of Internet scammers employing come-and-marry strategies and other schemes of social persuasion involved stylistically elaborate efforts at online self-presentation completely recasting bodily form—by swapping genders, altering race, and pursuing other manipulations of what are usually unalterable cues in face-to-face encounters. Such approaches leveraged more fully and creatively the possibilities of the virtual for persuasive self-presentation. Through such strategies, Internet users demonstrated a savvy awareness of the art of appearances and tapped into dynamics and pathologies of alterity that infuse relations between the West and non-West.

The way gender swapping enters into Internet scamming and how young, male scammers explain and pursue this practice gets at the gender roles, ideals, and expectations ascribed to by these youth and that they presume to be universal. The practice of gender swapping online was analyzed extensively in the early scholarly literature on online environments, specifically in what was observed in the Internet's earliest text-based, multiplayer game environments.[10] The comparison of the divergent motives of Americans who did gender swapping in that context versus Ghanaian Internet scammers is illuminating. In these early role-playing environments, participants pursued gender swapping as a way of seeking sexual excitement, to reexamine their own real-world sexuality and gender identity, or to generate empathic understanding of the other gender (Turkle 1995; Bassett 1997; Roberts and Parks 2001; Sundén 2002). However, mere identity exploration was not at all the appeal of gender swapping for young Ghanaians. This divergence indicates something of how the meanings and uses of "the virtual" vary across real-world geographies.[11] The online practices of Ghanaian youth run counter to the notion of a singular and emergent culture of the virtual or of a global cyberculture. The comparison of these examples indicates instead hybridity between offline and online cultures. Distinctive threads of cybercultural influence are forged from this online intermingling of culturally diverse perspectives, the materiality of the network, and the offline cultural norms that young Ghanaians draw from as a resource.

In the early theorizing around virtuality, gender swapping and other seemingly novel possibilities for fluid identity reinvention were treated as evidence of the social value of the Internet in its capacity for disembodiment.[12] The Internet came to represent the possibility of escape from being narrowly perceived according to one's visible and inflexible markings of gender, race, age, and so on. In online spaces, in contrast to the physical world, the body had to be constantly and willfully constituted. Scholars pointed to evidence in online role-playing environments where some users went still further, constructing bodies that had no physical world equivalents but were uniquely gendered, cyborg, or hybridized bodies combining human and animal or other nonhuman features (Reid 1996; McRae 1997). Elsewhere on the Internet, cerebral text-based discussion groups where bodily representation was shed entirely in favor of pure argumentation were similarly seen as another example of this transcendence of the body. The more eccentric among virtual reality theorists went as far as to propose a coming age when communication would happen through the purity of immersive, shared experience. They suggested that flexible virtual spaces would one day provide ways to transmit a "universal language" (Sherman and Judkins 1992) that would usher in an era of postsymbolic communication (Lanier and Biocca 1992). A global Internet, once infrastructure and access were put in place, was thought to be in this way inherently egalitarian. For users from political, economic, and cultural peripheries such as these youth in Ghana the expectation following this line of thinking was that social navigation online would be less problematic in certain ways than it was in offline worlds.

We may consider in more detail the way gender was specifically and mindfully performed by young Ghanaian men in practices of gender swapping to measure against such notions of the Internet's liberatory possibilities. A twenty-two-year-old named Gabby was the one I came to know the best among the scammers I met and interviewed. I visited the detached "small boys'" quarters where he lived at the front of his uncle's home and where he also set up shop as a barber to scrape together a small income. I interviewed him several times, sat with him at the Internet café, and went with him to a Muslim mallam in the Nima neighborhood of Accra to see about acquiring fetish to use on a chat partner. He described his personal history in detail, how his father had left and his mother had died when he was young; how his relationship with his uncle, his guardian, had become strained by mistrust, leaving him with a sense of great vulnerability about his future. He ultimately aspired to take an IT training course at NIIT, a multinational tertiary training school headquartered in India that had a local branch in Accra. His scamming activities, he claimed, were an effort to raise funds for this purpose. He got the idea to pursue scamming from some friends noting that "they were gaining a lot out of that . . . sometimes I would accompany them to the banks for the money. Sometimes one would go for 50 million cedis,[13] some would go for 100 million cedis. You know, definitely whoever you are, you want to get into that." The effectiveness of the come-and-marry scam in particular rested on his understanding of what is universal to intimate male-female relations. He stated confidently, "What wouldn't you do for your girlfriend? And so if I ask for money he will give it to me, definitely." Later in the conversation he noted that as a boyfriend "you have to be sending regular monies." When we met, Gabby's most promising online relationship was with a Chinese man who had proven himself to be dependable and "serious." Gabby was certain also that the man was rich. He had characterized himself as an international business man who traveled all over the world. It seemed from all indicators that a money transfer from the man was imminent. Over a period of five months I watched Gabby's scam strategies change and evolve though he did not ever manage to make any gains from the Chinese man or any other scam targets in that time.

One day in April 2005, I was with Gabby when he sat down at a computer in an Internet café in La Paz called EasyNet, entered a Yahoo! chat room and announced, "I AM ELIZABETH GRANT HERE TO TALK TO YOU

ALL," adding, "I AM 26 YEARS OLD." So began a session of attempted seduction in a gender-swapped body. Gabby stated that he'd typed his greeting in all caps to make it more visible, to gain attention over the other competing messages in the chat room. The chat room experience was a chaotic one, with the screen filling quickly with random comments, abbreviated text-talk, come-ons, insults, obscenities, and non sequiturs in different colors and fonts, with any single thread of conversation very difficult to follow. In practice, this room merely served as a jumping off point for one-to-one conversations with the people randomly encountered there. Once he appeared in the chat room, Gabby as Elizabeth was greeted more pointedly when a separate private window popped open on his screen with a succinct message: "hi." The first greeting was from someone claiming to be from Egypt asking Elizabeth if she spoke Arabic and then a second appeared from a person in South Africa. Next, perhaps in an effort to get more decidedly off the African continent Gabby moved to the "San Francisco" chat room, a space signified geographically but not actually materially bounded in any way by this label. He responded to approaches from strangers in this room stating mysteriously and invitingly, "'til you talk to me you will not know me." One person asked where he was located and when he told them truthfully Ghana he or she responded, "to [*sic*] far for me to meet." Many chat partners requested pictures from Gabby. He obliged by sending a posed (nonpornographic) studio photo of an attractive black woman that he had obtained from the site blackcuties.com. This image was meant to serve as an appealing as well as convincingly real visual representation of his online persona. What made photos from blackcuties.com especially well-suited for this purpose is that they could be downloaded in sets with multiple photos of the same model posing in a number of different outfits and settings. Gabby thought this enhanced believability by allowing him to ration out different photos of the same woman over time in the context of a developing chat partner relationship.

After several brief exchanges, Gabby ended up in a more extended conversation with an American going by the online name "EddieOK."[14] This was launched by an exchange of photos via email. The photo of EddieOK that arrived in Gabby's in-box was of a middle-aged, white man straddling a jet ski. He wore a t-shirt with a large American flag printed on it—a snapshot of an American middle-class or working-class archetype of patriotism and 'the good life" juxtaposed. EddieOK asked Elizabeth what she thought of the photo and Gabby, employing his best attempt at casual flattery and flirtation responded, "you look like, I don't know, you look like I want to be with you." In response, EddieOK evoked race directly for the first time in this encounter asking, "you like white guys?" These unfolding encounters illustrate an insistent anchoring of the conversation in the physical world. The importance of place was evident in the way it was introduced immediately on meeting or reflected in the dismissal "to [*sic*] far for me to meet." The tide of conversation pulled progressively from the starting point of a digitally coded and fleeting encounter to something more concrete, toward an increasingly high-fidelity and more authentic-seeming vision of this chat partner as realized through the introduction of pictures and places.

As observed in the chat room encounters between Gabby/Elizabeth and EddieOK, it was standard practice to move quickly toward a sense of the genuine, real-world person as the foundation of intelligible interaction. The common question "asl?" meaning "age, sex, location?" that begins almost every conversation in Yahoo!'s chat rooms and the popularity of Webcams in these virtual spaces is a reminder that many Internet users around the world remain very interested in bodies and locations. Anthropologist Michael Taussig has noted a more general human pattern in the way apparent, external forms, whether of dressed bodies or talismans, form a mutual relationship with the force of contact, connection, or communication within. And yet this relationship, he suggests, is an elusive one. To depict and to cultivate an appearance is a fundamental cultural compulsion. The force of the underlying connection relies on it rather than being simply hampered by it (Taussig 1993). Thus the idealization of postsymbolic or disembodied interaction overestimates the willingness and capacity of humans to dispense with such fundamental practices when they go online.

The original trope of liberatory disembodiment on the Internet rested on a certain notion about the nature of human social behavior and, in particular, ideas about what conditions incite conflict or facilitate communion. The idealization of immediate, disembodied communication suggests that human conflict arises from not seeing one another transparently enough. To adopt this notion of conflict was to accept no legitimate or fundamental disagreements or differences between people, rather, that our visible exteriors (indicating race, gender, class, etc.) are meaningless and misleading distractions implying differences that do not truly exist.[15] However, the breakdowns and misunderstandings between Ghanaian youth and foreigners online suggests something more fundamentally embedded in communication that makes differences apparent and that can generate offense and avoidance. Thus merely concealing identity markers does not pave the way for smooth and unprejudiced communication.

The attempt by the young Internet scammer Gabby to present himself as a woman online indicates that the impulse toward identity manipulation and gender swapping is not exclusively a Western preoccupation. However, practices of gender swapping among young Internet users in Accra clearly were motivated by a different interpretation of virtual possibilities. Multiple ironies and a deep sense of ambivalence took the place of the earnestness of utopian visions of cyberspace as a place for unconstrained being. The capacity for disembodiment was understood among young Ghanaians as principally facilitating flexible self-presentation for the sake of persuasion and this was found to be especially useful in forms of Internet scamming. Authenticity in this context was a matter of great concern among scammers who had to convince scam targets that their online appearance was also their offline reality. Gabby reported being quizzed by skeptical strangers in tests of authenticity—asked, for example, his bra size as a trick to ferret out whether he was a man posing as a woman.[16]

What was also apparent in the strategies of scammers was an understanding of bodily representation not as an obstruction to be rid of but as a resource. The flexibility of the online realm was appealing not for the capacity to become disembodied but for the ability to be flexibly rebodied. Male scammers saw an opportunity to garner attention and obtain favors from foreign men through the seductive qualities of a shapely, female form. They favored narratives embedded in a sometimes elaborate but specific sociopolitical context with real-world referents. Though ultimately fictional, the narrators in these tales had an identifiable race and gender as well as a specific nationality, family ties, and professional affiliations. The act of gender swapping as a scam strategy demonstrates how embodied identity prevails online as an enhancement to the believability and persuasiveness of the scammers' appeal for aid.

Manipulating Representations of Africa for the Foreign Gaze

Beyond the gender-swapping strategies of scammers, which more generically rest on young male Ghanaians' ideas of universal gender norms, some of the more intriguing strategies drew on scammers' savvy insights into Western representations of Africa and Africans. What they referred to as the "format" of the scam evolved over the time period from 2004 to 2010. When once they posed as exotic, Ghanaian women who were available to and interested in older foreign men (as the previous section described), many scammers, recognizing the suspicion they were increasingly generating by identifying as African, were by 2010 starting to pose often using a white persona, sometimes as a man, of different ages, whatever matched to the targets' interest and was most likely to evade suspicion. They also were more broadly targeting various dating subcultures—such as overweight, older "silver singles," or gay and lesbian dating sites. By 2010 a gain of a few hundred or even a thousand dollars appeared possible even among the less experienced, free agent Internet scammers employing these adjustments to their narrative and self-presentation.

This second layer of strategy among Internet scammers involved the use of narratives of corrupt governments, shady business deals, religious piety, illness, war, poverty, untimely accidents, and natural disasters. Such a repertoire of representations calls out for an analysis beyond the implied cultural relativism and the agnosticism about power that the tales of cultural mismatch and misunderstanding so far considered suggest. Instead a postcolonial reframing must be brought in to further understand circumstances of Internet use in Accra. Discussing these narratives with scammers and nonscammers alike in Ghana usefully illuminated contested issues around the relationship between Ghana and the West via the struggles and strategies of digital depicting. Scammers were self-aware mimics. They specialized in portraying Western archetypes of the African Other. The narratives scammers produced were emulations of themselves as Africans but as viewed through the warped self-interest of Western representational practice. Scammers stepped out of themselves to imagine the foreigners' "gaze" and sought to meet the expectations of that gaze in a way that gave them a strategic advantage. This was not simply about how the Western gaze functioned as a constraint but also about "the interlocking of gazes" as Nyamnjoh and Page detail in studying characterizations (or caricatures) of white people among Cameroonian students (Nyamnjoh and Page 2002, 609). Internet café users were also, at times, tripped up by their own limited imaginings and perceptions of Westerners.

Ghanaians were widely aware of the way the foreign media typically represents Africa as homogenously war-torn, poverty stricken, and chaotic. This was a source of frustration among many who felt these characterizations were in many ways inaccurate in relation to Ghana. The country's status as a Heavily Indebted Poor Country (HIPC)[17] was also a source of shame and was sometimes publicly lamented in church services and in the local media. To be deemed heavily indebted and poor and therefore in need of special foreign aid and debt relief was a threat to many Ghanaians' sense of agency and dignity. This sensitivity about foreign representation meant that friends and strangers alike at times tried to dissuade me from taking photos of scenes that might suggest poverty. They demonstrated an awareness of the Orientalist tendencies of foreigners to depict their society as passive, poor, and strange through imagery (Said 1978). Challenging the way I, as a foreigner, framed photos was one way in which people in Ghana were able to effect control over their representation to the outside world, but beyond this they had few other options for addressing the foreign representations that were imposed on them.

Foreigners' representations of Africa in chat room conversations often mirrored images from the Western media. Ghanaian Internet users found that their foreign chat partners were quite ignorant about Ghana and Africa more generally. Some chat partners expressed fear over visiting Africa, a fear of violence or disease. Kwaku, the young videographer, described to me his bemusement when a Palestinian chat partner expressed the fear that he might die if he came to visit him in Ghana because of the wars going on in Africa, somehow looming as a larger threat than the conflict he was living through in his own homeland. Gabby, before his days posing as Elizabeth Grant, back when he was a schoolboy, sought out pen pals through the postal mail telling his letter recipients, "My parents are dead, please help me, I'm in Africa, you know the situation in Africa." Justifying such a strategy he lamented, "You see Africa is never printed as it is out of the continent Africa . . . if you are not in Africa all the pictures you see in Africa are diseases. . . . These nice, nice places will not be broadcasted. . . ." Through exposure to media representations and more directly through interactions with foreigners online and in person, Ghanaians were forced to confront how they were constructed by Westerners as the Other. Such a portrait is constructed in terms of what the viewer has that the viewed lacks and is built on anxiety and a desire for self-affirmation. The Other is defined by absence. Internet scammers similarly constructed Westerners as Other in developing their online personas, but using a form of *double consciousness,* a term W.E.B. Du Bois defines as "this sense of always looking at one's self through the eyes of others, of measuring one's soul by the tape of a world that looks on in amused contempt and pity" (Bois 2003 [1903], 9). Westerners are conversely described in terms of what they have that Africans lack. The Western Other is therefore an inversion of the African Other. As Africans are typified in the West by poverty-stricken famine victims on the television, Westerners are typified by wealthy celebrities. Whereas Africans are needy, Westerners are greedy. This mutually agreed-on asymmetry is exaggerated into mutual misunderstanding.

Young Internet scammers believed these foreign perceptions of Africa could be usefully manipulated perhaps even in a way that served a subversive sense of justice. In this way a kind of marginalization brought on by the perceptions of outsiders could be used pragmatically as a tool for extracting money. As young Ghanaians imagined themselves as Other, they found a limited set of archetypal identities they could perform that would be treated as believable and sympathetic by foreigners. They believed that their desires for the capital needed to start a legitimate business or to go to an IT school were not persuasive enough to grab the attention of a foreign contact, as Gabby noted (as quoted previously), "if I put my real picture or if I tell you this . . . my family background, my status, my financial background you will not even want to talk to me." Scammers came to understand the foreign gaze as one that expected an asymmetry, expected and responded to them as needy, but not as potential partners. It expected medical bills, not IT schools. In response, scammers created alternate identities that catered to the perceived prejudices of their foreign chat partners. They performed as a needy African orphan, as an attractive African woman seeking rescue, as a participant or victim of a corrupt African government regime, or as a God-fearing Christian pastor seeking funds to help improve his impoverished community. This was mimicry directed back at the West as a provocation, an effort to manipulate from afar.

The repertoire of strategies employed by scammers also included mimicry of Westerners as a technique of persuasion in line with the compulsion noted by Benjamin "to get hold of something by means of its likeness" (Benjamin 2001 [1936], 52). To "get hold," for the scammer, was first about gaining attention, that increasingly scarce commodity on the Internet, and then about effecting a kind of control over distant others, to make them responsive to their suggestions and appeals. For example, after interviewing a young self-proclaimed pastor, I was approached by an attendant from the Internet café where I'd recruited the young pastor. This café attendant wanted me to understand who this pastor really was, as alternate to his self-presentation. He handed me a scam letter ostensibly created by this young pastor. It was written in the personae of an elderly, white, American woman to appeal on his behalf to an American televangelist and pleaded:

> Dear Paul Crouch I am sending this to ya begging ya to help this man of God in Ghana Africa I found him on the Net and he adopted me as his mom but w/ tears in my eyes I am an elderly woman on s security and have no means to do what's necessary for him . . . he do0es not have food/shelter amnd money and sleeps at times at the bus station and park as he is "black" and no one gives him aid and cannot get a job of this reason . . . [I am a white woman from the south] God blass ya !! [name and address omitted]

In the letter the young pastor constructs an embodiment abruptly and parenthetically in terms of race, gender, age, and regional location. His third-person narrator is positioned as a sympathetic advocate. In his capacity to select any bodily representation he desired, he chose a narrator of the same religion, nationality, and race as the televangelist (though of a perhaps more trustworthy and sympathy-inducing age and gender). This scammer illustrated an extensive awareness of not only Western media representations of Africa, but also representations of the African Diaspora (in his incongruent and confusing reference to being unemployable in Ghana because he is black) and American regional subcultures (in his implicit tie between religious devotion and the South). What he formed, in the end, was a merger (albeit slightly convoluted) of these different domains.

Constructing the scam strategy involved ascertaining and playing on a certain characterization of the Western Other but in trying to justify such morally questionable acts scammers drew on additional characterizations. Yaw, who claimed to be in touch with a number of very successful scammers operating out of a large, centrally located Internet café, told stories that cast Westerners in a darker light, not as ignorant people susceptible to the influence of their biased media but as blatantly prejudiced and as actively disadvantaging Ghanaians and other Africans and therefore deserving of the Internet crimes committed against them. Yaw noted that his friends didn't feel bad about scamming because, he noted, "they say the white man is the biggest thief." Yaw also pointed to feelings of resentment over the harsh restrictions on migration (as considered in chapter 2) that prevent young Ghanaians from seeking better opportunities overseas, noting that "because they can't go to America they will take money from Americans." He pointedly asserted that Westerners perceive Africans as not clever and capable enough to carry out a scam. They play on narrow and unflattering perceptions of Africa and as Apter notes a more general sense "that the 'third world' plays by flexible rules" (Apter 1999, 271). In a satisfying sense of revenge against this perceived racism, in the ideal scam it is precisely through this prejudice that these Westerners are taken advantage of. Scammers and their sympathizers imagine a moment of realization when racist scam victims must confront the fact that they have been outwitted by individuals they had thought to be inferior.

Resentment among scammers and other Internet users in Ghana around the inauthenticity of online self-presentation stemmed from being limited to a set of false perceptions and distorted archetypes that they viewed as alien. In this way they were operating with what de Certeau defines as a tactic. A tactic is the strategic work done by "the weak" who lack a space of their own from which to relate to what is external. The lack of space is determined when external forces refuse to recognize that position as they define it for themselves. This forces "the weak" to relocate. A tactic "insinuates itself into the other's place" (de Certeau 1984, xix). Internet scammers were doing this in a material and discursive space. As de Certeau notes, "the weak must continually turn to their own ends forces alien to them" (de Certeau 1984, xix). Through tactics Internet scammers sought to subvert and transcend a disadvantageous position within society and the world using the very representations of Africa and Africans defined apart from and against them.

Conclusion

The analysis of young Ghanaians online in this chapter reworked an existing literature on the "the virtual" and its concern especially with Western postmodern and poststructural theories of identity (Poster 1995a; Slater 1998). Early in its history, the Internet was seized on by scholars as a way to interrogate such theory empirically: by observing how users extended themselves into these new online spaces. The parallel world of cyberspace, it was then thought, offered the novel capability of immediate, interactive communication combined with an ability to shed bodily form and to represent the self purely through text. In utopian imaginings, such a loosening of the material represented liberatory possibilities.

Given the ever-increasing ordinariness of the Internet in the West (and weak delivery on the promise of virtual reality technologies emerging around the same time) such thinking is quickly fading just as the notion that the telegraph would yield an era of cross-cultural understanding now seems quaint (Marvin 1988). However, a more recently emerging rhetoric in the international aid sector recycles many such themes championing the possibilities the Internet offers for cultural exchange and mutual understanding within the Global South and in bridging developing and developed nations.[18] Thus the somewhat dated thread of cyber-utopianism is worth revisiting with fresh empirical insights. Populations in certain online spaces have become more truly diverse and global—putting such ideals to the test. For young Ghanaians, rather than enjoying an unencumbered being through the Internet as a space for play and richer self-understanding, they came to see it as a space of persuasion and for enrolling allies. The possibilities of virtual inhabiting came to be understood by these youth in the way they were situated (or wished to be) vis-à-vis the West.

Young Ghanaians' strategies of online interaction underlined the immense challenges of attracting an audience and maneuvering these connections from a great distance. Yet the complicity of Internet scammers in employing stereotypes of Africa and Africans to pursue scams ironically had the effect of reinforcing and reproducing the representations they most resented. Harald Prins examined a similar form of complicity in an analysis of what he terms the *primitivist perplex,* when indigenous groups make use of outdated and inaccurate portraits of a primitive and nature-centered way of life in order to make appeals to the larger dominating society for autonomy and preservation (Prins 2002). Such willful and knowing misrepresentations served pragmatic purposes. Marginalized groups create and employ them in an attempt to secure material gains such as money, land ownership, or government resources.

In the end, the potential for the empowerment of marginalized groups through new media technologies relies on producers who represent themselves in authentic as well as strategic ways. The problem of authenticity and its seeming incompatibility with persuasive self-performance was significant to Accra's nonelite Internet users. Empowerment, as Faye Ginsburg argues, means ideally "rendering visible indigenous cultural and historical realities to themselves and the broader societies who have stereotyped and denied them" (Ginsburg 1994, 378). This means knowing how to speak to these broader societies to gain attention and inspire action. The question therefore is what conditions facilitate empowering self-expression and what conversely provokes individuals to become complicit in their own damaging self-representations? In the reproduction of Western archetypes of the African Other by scammers and in the low trigger point among foreign chat partners for avoiding and blocking young Ghanaians, these broader realities of the daily life of young urban Ghanaian remained largely invisible. In this way the opportunity for cross-cultural exchange and understanding and for true partnership was often thwarted.

4 Rumor and the Morality of the Internet

There is something really about this Internet, there is something that is really making my friends rich.
—Gabby, twenty-two-year-old Internet café user

To arrive at a coherent reading of a new technological form, users may draw from different kinds of scripts. As scholars have argued, advertising—well-recognized as a kind of script—intentionally plays into this process by depicting new technologies in a way that is meant to shift public understanding toward a market-desirable, gender-normative, and otherwise reassuring or stabilizing role for the technology in the social order (Nakamura 2002; Mackay and Gillespie 1992; Lally 2002; Hubak 1996). Scholars taking a more materialist analytical stance suggest that the technological form itself (here the computer, monitor, peripherals, the operating system, and other software) as written by designers and engineers offers the primary and definitive script to be read by users (Woolgar 1991; Akrich 1992). This particular materialist argument and the way it treats sense-making as coextensive with direct material manipulation of the technology is specifically called into question in this chapter and the next. Looking further afield at spaces where a technology is handled indirectly, I find that formal message circulations such as advertising are not the limit of what we might consider as scripts. In Accra specifically, several new genres came to the fore, a result of different dominant cultural institutions (e.g., churches) and the nature of the early, underground adoption of the Internet among youth in the Internet cafés. Such circumstances of use promoted more informal scripts. Rumor is one of these genres and is the focus of this chapter.

Among Internet users in Accra, Ghana rumors often supplanted direct human-machine interaction as the privileged mode for understanding how the Internet works. Rumor amplified rare instances of wildly successful or especially harmful encounters with the Internet. This was particularly apparent in discussions of Internet fraud and scamming activities that in 2005 appeared to be much more widespread as a phenomenon of speech and social imagination than as one of firsthand experience. Many young Internet users admitted that they themselves had not yet made any money on the Internet but expressed the conviction that it was being widely used by clever, young Ghanaians (and others) to acquire thousands of dollars, the latest fashion, or the newest technologies. Through the retelling of rumors, Internet users collectively re-envisioned the technology as a tool for making "big gains." Rumor fed the local circulation of meaning for managing the novelty and especially the moral uncertainty of the Internet.

Rumors are taken seriously in the analysis that follows, treated not as false and fantastical tales but as convincing and durable speech acts that had consequences for how the Internet ultimately came to be materialized in Accra. Although popular understandings of rumor often emphasize their questionable veracity,[1] they are more aptly defined, I assert, as secondhand accounts. Rumors recount what is believed by the teller to be a real-world happening though one that was not directly experienced. Such a definition captures the social role and continual and widespread circulation that marks rumor as a distinctive oral form. In the examination of rumors about Internet scamming we gain the opportunity to peer more closely into the reception and interpretation of the Internet as shaped within the routines of other cultural formations in urban Ghana, not just in the context of what unfolds inside the Internet café and while users are seated at the computer interface. Rumor is part of a broader domain of informal media or what anthropologist Debra Spitulnik refers to as "small media," a category that includes such forms as jokes, graffiti, self-published texts such as pamphlets, and other similar formats and genres of communication (Spitulnik 2002a). Rumor has been shown to play a particular role in African settings. Spitulnik argues that under state-owned media monopolies and conditions of media repression that have historically existed in many African states, rumor and other informal modes of message dissemination offered a relatively safe channel for ordinary citizens to produce, consume, and circulate messages. In the context of urban popular culture in Cameroon, Mbembe recognizes rumor as a form of resistance among the repressed, a "poor person's bomb" (Mbembe 2001, 158). What is applicable to the current case is the way the lateral (rather than hierarchical) message circulation characterizing rumors was accessible to marginalized populations such as these urban youth who lacked the resources or skills to pursue more formal message circulation. In relation specifically to technological sense-making, rumor circulations offered a way of propagating interpretations of technology among users and thus constituted a process of collective (rather than individual) sense-making. Through such means, users produced and materialized the technology for one another.

The analysis of rumors contributes a significant extension to the material-semiotic accounting attempted in this book. In particular, the remarkable durability of this oral form inverts a taken for granted ordering in such forms of accounting that tends to privilege manipulations of the object world over other modes of practice. In the relocation of semiotic reasoning from words and language (traditional semiotics) to object and actor relations (material semiotics), words and the work of representation have been left with a rather ambiguous role. Speech, when it has been considered, has often been held apart as a uniform mode of action that generates only weak effects.[2] Law, for example, comments that "voices don't last for long and they don't travel very far. If social ordering depended on voices alone it would be a very local affair" (Law 1994, 102). However, the importance Ghanaians placed on words and their potency, particularly the significance of charismatically performed, persuasive speech, necessitates a second look at the role of what is spoken in the performance of material-semiotic ordering and how it shapes the way a technology such as the Internet is engaged and enacted. Certainly it is a mistake to treat speech as wholly immaterial (in comparison to object manipulations) because it inevitably requires the materiality of the vocalizing organs, bodily message production, and human memory capacities. What emerges from more careful observation is that speech acts vary quite dramatically as far as their ephemerality and potency. Rumor, in particular, is a format that is especially durable.

Rumors about the Internet in Ghana in their telling and retelling came to constitute a persistent social imaginary. As defined by the philosopher Charles Taylor, a social imaginary is "the way people imagine their social existence, how they fit together with others, how things go on between them and their fellows, the expectations that are normally met, and the deeper normative notions and images that underlie these expectations" and is, as he notes, "carried in images, stories, and legends" and that "makes possible common practices and a widely shared sense of legitimacy" (Taylor 2002, 106). In Ghana's Internet café scene this specifically came into play in managing the new moral order that had come to include distant and foreign actors. The very real presence of this particular social imaginary (rematerialized each time such a rumor is retold) was important to the equilibrium of Internet activities that without such compelling tales may very well have been abandoned by users. These rumors provided a set of characters, actions, and a sequencing of steps setting forth a map or script for effective use. What was done at the interface (and thus the materialization of the Internet itself through the activities of users) was also shaped and informed by these tales. Rumors smoothed over the delays experienced by users in realizing gains from the Internet. In this way such tales filled the gap in time between immediate experience and expected outcomes reconciling some of the uncertainty of Internet use by offering rumored gains as a temporary substitute for realized ones.

In rumors about Internet scamming, users not only managed questions of efficacy but also issues of morality, opportunity, and aspiration that seemed altered and newly challenged by the technology. Among all forms of Internet use in Accra, Internet scamming had the highest stakes. It was an activity that held the promise of unlimited financial opportunity. At the same time it represented a dangerous entanglement that could damage the morality and finances of those involved. The idea that these crimes were being committed by their peers cast suspicion on young Internet users and contradicted the positive representation they sought to portray of themselves and their society. The relationships depicted in rumors helped to motivate certain uses of the Internet and established a sense of morality and safety. Rumor played multiple roles. It did not simply provide for information transfer as certain theories of rumor once suggested (Shibutani 1966). Ways of effectively using the Internet as defined in rumors were shaped by the pressure of establishing a kind of relational stability between Ghanaian youth and the heterogeneous populations they encountered online composed of potential friends, patrons, victims, or victimizers.

Rumors as Accounts

In addition to the remarkable durability of rumor, the way these tales managed to supersede direct experience in the case of Internet activities in Accra also demands explanation. What makes it possible for believability to skew toward such indirect and unverified tales over what these users saw and experienced for themselves? Why isn't it the case here that seeing is believing? Historian Luise White offers one explanation in her study of rumors about colonial vampires in East Africa. Such rumors told throughout Kenya, Uganda, the Belgian Congo, and elsewhere in the region implicated institutions of the colonial state—the fire brigade, the police, hospitals, and medical practitioners—in the kidnapping and killing of African colonial subjects to drain and collect their blood (White 2000). Although European colonists who came across such rumors tended to treat them as evidence of the superstition and credulity of native African populations, White disputes this dismissal of such rumors as belief without evidence. Believability, she suggests, rested on alternate criteria among those colonial subject populations. It was the continual and widespread retelling of these stories, the fact that so many others reproduced and believed them that came to be taken as evidence of their truth. As one of her interviewees noted of one tale, "it was a true story because it was known by many people and many people talked about it . . . they would not talk about it if it was not true" (White 2000, 31). This perspective was also reflected in the explanations of my own informants as well, such as Kwaku, an Internet café user in Mamobi, who noted of one such local tale, "the thing is I heard it, but it is true . . . because people are witness. It was not one person who said, it was a lot of people who were saying it." What rumors offered was a coherence that was lacking in the ambiguities and confusion of direct sensory perception. They yielded "ideas on which experience can be based" and thus functioned as "social 'truths'" that were compelling and reliable (White 2000, 33). The reverse question may also be reasonably asked—why should we assume that the narrowness of one single person's direct experiences with the Internet would necessarily be more definitive than such broader social truths?

In search of further explanation for the apparent belief in secondhand accounts over direct experience I draw some insight from the way accounts and the actions accounted for are distinguished in the theoretical contributions of ethnomethodology. Accounts have been examined ethnomethodologically in many forms including medical record keeping (Garfinkel 1967), instructions for using a photocopier and computer code more generally (Suchman 2007; Latour 1992), scientific papers (Latour 1987), and even answers to nosy questions posed by anthropologists (Bourdieu 1977, 17–18). What such an analysis has argued is that accounts are neither determinants of action nor a comprehensive post hoc representation but are a resource that sustains action and comes into play in resolving breakdowns in the smooth flow of unfolding activity. One recurrent theme is the inevitable incompleteness of accounts. They are an ideal or generalized case denuded of the exceptions and contingencies that emerge in unfolding experience. The further an account is elaborated, the less flexible it becomes as a resource. This argument liberates accounts from the role of merely mirroring reality and shows their coexistence alongside the actions they depict making such practices possible.

Rumors may be considered as a particular kind of account. When rumor telling is reframed as accounting work the appearance of contradiction dissipates. Rumor and direct experience are no longer seen as competing modes but rather intrinsically connected components of a singular practice of Internet adoption and use. According to such a schema it is possible to see instead how the continual retelling of rumors about big gains made the actions these users carried out at the computer interface defensible (publicly and personally) and thereby enabled their continuation. Understanding rumors as accounts also helps to avoid the tendency to divorce the symbolic (the realm of words and narratives such as rumor) from the material as two distinct worlds operating in parallel. Rumors about big gains were more than a window into a representational realm for the observant researcher to understand better how people think about the technology. They were a product not only of the interview event (though it was in interviews that I encountered these tales) but also of the social world. It is possible to make this claim because of the way rumors reference a source (the news, a friend, or more ambiguously "I heard") indicating a separate place and time where the rumor was received. The patterns in retelling, the recurring images and reference points, indicate further the social existence of these tales beyond the idiosyncrasies of individual Internet users.

Rumors as accounts, however, diverge in several interesting ways from the sorts of accounts typically considered from an ethnomethodological perspective. Generally such accounts are produced by actors about their own actions or directly observed actions. By contrast, rumors are accounts claimed to originate in real-world events but that have not been witnessed directly by the teller. As a consequence, not only does rumor diverge from an impossible ideal of transparent representation for all of the reasons noted, but also due to the alterations introduced by the whole unknown chain of individuals involved in its transmission. Furthermore, rumors (often taking on the form of morality tales) bring the socioaffective demands of technology use to the fore in a way standard ethnomethodological accounts do not. In this way, rumors also contribute to the motivation to use (or not use) a technology in the first place as it relates to a sense of the moral self. The voluntary nature of Internet use in Ghana makes such questions of initiation and motivation central to explaining what the Internet became and its particular user population there.

In ethnomethodological approaches there is a tight coupling between actions and accounts of those actions as circumscribed by the borders of the micro world under consideration (e.g., the health clinic). Rumors, however, do not submit to such bounding. In the broader literature on popular culture and rumor in African studies, the fantastical and imaginative is an integral part of the role rumors play in sense-making. Such tales engage with the metaphorical and mythological, the realm of contemporary legend in popular belief. Furthermore, in the contemporary era of highly connected and media-infused societies, the realm of popular creative expression draws increasingly on imagery and ideas from far-flung sources rather than originating in the immediate milieu (Appadurai 1996). In line with this shift, the rumors about scamming I heard in Accra were distinctive for their global span. As will be examined in more detail in the following section, these rumors sometimes featured celebrities such as Oprah Winfrey and Michael Owens, the British soccer star, envisioning what their lifestyles are like. The stories unfolding in these rumors took place, in part, in foreign countries (one that the storyteller had never visited) and referenced foreign institutions such as the CIA and the global credit system conjecturing about the inner workings of these entities. These circulating rumors about the Internet are evidence of what Appadurai has described as a global "rupture" yielding a new status for the work of imagination that has "broken out of the special expressive space of art, myth, and ritual and has now become a part of the quotidian mental work of ordinary people in many societies" (Appadurai 1996, 5). As a result, accounting work and creative expression are becoming increasingly indistinct. Imaginative outpourings are woven into the flow of everyday decision making rather than being a separate space of cathartic release or professionalized artistic practice. The scope of resources to draw from in accounting work becomes maximally broad and flexible, though one might also say, increasingly distant from experienced reality.

A Typology of Rumors about the Internet in Urban Ghana

In the corpus of rumors collected and analyzed by scholars, there are certain recurring themes in subject matter and plotline. One subset is those rumors that comment on relationships between the powerful and the disempowered. In Patricia Turner's book on rumor in African-American culture, she notes that such tales of power are often rendered on the body in forms of corporal control or corporal punishment (Turner 1993). They include stories about the systematic sterilization of undesirable minority groups—by faceless corporate entities or by hostile foreign governments (Feldman-Savelsberg, Ndonko, and Schmidt-Ehry 2000) or by one's own government (Kelman 2009). Similarly corporal are stories of cannibalism and of vampires as White analyzes (White 2000). Following this theme, certain rumors I heard while doing fieldwork in Accra (that are beyond the scope of this chapter) reveal a sense of vulnerability and underlying mistrust between insiders and outsiders, locals and foreigners. A mistrust of authority and thus of "authoritative" sources is a further circumstance underlining the believability of rumors stemming from their lateral, populist circulations. In Ghana, there was the story I was told about a Ghanaian man who discovered a cure for AIDS but who was killed (by an unspecified assailant) for fear that the multitude of NGO workers involved in AIDS awareness and relief would lose their jobs. Other rumors emerged in the aftermath of the death of an American student studying at University of Ghana at Legon of cerebral malaria. I heard two variations recounting this death in rumor. One version (circulating among young Ghanaians at the university) was that excessive pot smoking had rendered him insensible and that he had neglected to seek medical care as a result. Interestingly, an opposing version circulating among the expatriate crowd was that the young man had been denied care at a series of clinics and hospitals, his friends expelled for being rude to the hospital staff. Other rumors dealt with vulnerabilities of a more local nature. For example, I heard about the magical powers of certain criminals in the Nima neighborhood, that they were immune to bullets, could smell money, and could become invisible.

The substance and subject matter of rumors specifically about the Internet reflect the subjectivities of youth emerging in a new era of connectivity that is accompanied also by certain anxieties. At the nexus of this anxiety was the morally questionable activity of Internet scamming, a topic that many rumors about the Internet addressed and attempted to reconcile. Rumors on this subject matter existed in several forms in Accra. Many were told as "success stories" emphasizing the prosperous outcomes of these activities, the enhancement of personal agency that came through contact with the Internet. These were typically told by young men who had attempted Internet fraud or scams themselves or who sympathized with such activities. Other rumors about Internet scamming were told as tales of deceit highlighting the scamming process and the misrepresentations scammers carry out in their forays online. Those who told deceit rumors often contextualized them within stories about defending foreigners from scams. This served as a way to position themselves as affiliated with foreigners in a protective capacity and playing the role of cultural insider and guide. Rumors were also told as victimization stories. These sorts of tales were offered by men and women and described local rather than foreign victims revealing a greater sense of vulnerability. Victimization stories more directly questioned the morality of such activities. In these various modes young Ghanaians described Internet crime as an attractive activity or as an activity that they feared or lamented. However, all rumors about Internet crime took as a given that Internet crime activities were widespread and successful.

"Success stories" tended to be the most elaborate Internet crime rumors and had a common structure. They started with the acquisition of a link to wealth located abroad. This was either a personified link (such as a boyfriend from a dating Web site) or a nonpersonified link such as a functioning credit card. The scammer/fraudster used the link to gain money or goods through Internet purchases or money transfers from the scam victim. In the narratives' resolution the scammer realized a total life transformation through his financial gains which provided a jumpstart to legitimate, sustainable, and prosperous adult status. Scammers disappeared from the scene after their big gain. "I think he lives abroad now" was a typical epilogue to a scamming story. Internet crime was never described as a permanent way of life. This type of story about gains acquired through the Internet served as a fantasy of rapid capital accumulation.

An example of a particularly elaborate scamming success story is the one I was told about a young man who gained $20,000 through a scam. The scammer was said to have posed as a woman on the Internet. "She" found an American, male chat partner whom she convinced to pay for her schooling in the United States. The scammer went as far as to apply for and gain acceptance to an American university as a woman. The chat partner paid the school fees directly to the school in the United States. Then the scammer canceled his enrollment and took the refunded money. He used half of it to buy a house and the other half to buy treasury bills using the interest to buy and sell "ladies clothes and shoes." Rumors were not always as detailed as the story about the $20,000 gain though. Many were offered with only the most basic of details.

Such narratives of success and deception depicted in rumors reflect how young Ghanaians perceived their position in the global economy. They illustrate a model of the world in which vast accumulations of wealth were located primarily overseas. An ability to advance in life is unlocked by gaining access to these financial accumulations. This access, in turn, comes only from establishing contact and persuading foreign gatekeepers to this wealth to form partnerships, whether through legitimate or criminal means. The tensions apparent within these rumors illuminate certain contested issues within Ghanaian society as following examples will illustrate.

Rumors and the Construction of a Moral Order

Within the structure of rumors the teller related subjects and objects to effect a stable social world. In particular, Internet crime rumors played a significant role in strategies for constructing desirable representations of self through affiliation or nonaffiliation with particular groups. Rumors could serve as both self-confirmation and as "impression management" (Goffman 1996 [1959], 208) for those who told them (Paine 1967). These rumors were used by Internet café users primarily to establish themselves as either good or effective (or ideally both) in relation to the Internet. These two qualities reflect a shared concern among Internet users with the new opportunities presented by the Internet as well as the moral trade-offs potentially required to pursue those opportunities. Through rumors Internet users cast characters, including themselves, in an unfolding drama around Internet crime. The following conversation among Gabby, an admitted scammer; Kwadjo, an ordinary, nonscamming Internet user; and me demonstrates the casting of characters:

Gabby: Sometimes the people whose credit cards are being used are these rich people . . . they don't really notice it, the money is reduced.

Kwadjo: Laptop to Bill Gates is a peanut.

Gabby: He will not see it. Last time I heard on the news that this lady, Oprah Winfrey, sixty million dollars was missing on the credit card or something like that whilst all the things the money was used for she didn't order them. And I don't know whether they arrested people who bought the things.

Kwadjo: Sixty million dollars!

Author: But are most Americans like Oprah or like Bill Gates?

Gabby: Eh hehh, well if I get the credit card of Bill Gates . . .

This conversation is an example of how foreign celebrities or sports stars were sometimes cast as scam victims who were impervious to harm. Similar to Gabby's comment about Oprah Winfrey, Daniel, a nineteen-year-old Internet café user, referred to famed UK sports star Michael Owen in his interview noting, "in Ghana here we have some people who are having . . . direct access to people's accounts. I learned some months ago, they've utilized the money in Mike Owens's account."[3] Invoking the names of famous celebrities in this way created distance between the criminal and victim of Internet crime. This distance created a sense of total nonidentification with scam victims who were represented in such rumors by celebrities who are maximally different given their fame, wealth, and nationality. As Gabby noted, "Sometimes the people whose credit cards are being used are these rich people." My appeal (as the owner of several credit cards) for a more representative understanding of scam victims in the comment, "but are most Americans like Oprah or like Bill Gates?" was totally ignored by Gabby who instead followed up by fantasizing about obtaining Bill Gates's credit card. Internet scammers in this way construct a scenario in which scammers steal from the very rich (who are unharmed by this act) and give to the poor, whom they define as themselves. For example Gabby, who had spent six months attempting to scam people online, claimed his deceitful activities were not greed but "because of circumstances that I'm doing it. Sometimes if I'm not really in need I wouldn't go and dupe somebody for money." Through these casting strategies rumors functioned to create a more morally sound relationship between young Internet café users in Accra and foreign victims of Internet crime.

Constructing nonidentification has historically been a strategy of moral justification. There is a precedent to the use of rumors to cast groups and individuals out of the human race entirely as either superhuman or subhuman. Celebrities, sports stars, and the fabulously wealthy represent something verging on superhuman. At the other extreme, as Turner notes, in the first encounter between English explorers and native Africans, both groups concluded that they had made contact with cannibals. For the colonialists who followed, this rumor was a strategy of attributing their greatest taboo to Africans, casting them as subhuman and served to justify dehumanizing exploitation of their bodies and their land (Turner 1993). In this way victims of exploitation or of crime are seen as already too depraved or too privileged to need or deserve a measure of human sympathy.

Just as Internet users cast Internet crime victims to put their peers in a relationship with the Internet that was good or morally sound, they also cast Internet crime perpetrators using similar strategies of nonidentification. This also served to maintain moral stability but in this instance through the identification of scapegoats. For Internet users this was as simple as ascribing Internet crime to maligned local minority groups. In Accra these were Nigerians, Liberians (often refugees), Muslims, and people living in certain crowded slums in Accra such as Nima, Mamobi, or Newtown. In interviews, Muslims living in Mamobi (one of the primary fieldsites for this research) expressed an awareness of their marginalized and maligned status particularly in relation to the Internet. However, they found ways to reinterpret this status so that it could serve as an advantage. In Mamobi, Internet users often confirmed a tendency toward Internet crime activities in the area but reframed it in a way to enhance their self-presentation strategies.

Mamobi is an area in central Accra that has historically been a destination for rural-urban migrants coming from the north of Ghana. The significant presence of minority groups in the area including Muslims, Nigerians, and people from smaller ethnic groups in the north of Ghana contributes to this marginalization. In addition, the area is very poor, densely populated, and has an inadequate infrastructure. The water taps were locked in the area necessitating long journeys to fetch water in the mornings. There were also major problems with sanitation, trouble with getting garbage picked up by the city, and a severe shortage of toilet facilities. Inhabitants of Mamobi referred to it as a *zongo,* sometimes drawing an analogy to the American urban term *ghetto*. Zongos such as Mamobi were seen as areas with high crime rates including armed robbery, mobile phone snatching, and Internet fraud. There was also a general sense in this community that the government and its representatives were neglecting their needs and concerns.

In this marginalization, however, inhabitants of the area were able to reframe their community's reputation to build an empowering narrative of self-sufficiency. Hamza, a young man from Nima, a slum adjacent to Mamobi, noted that as a result of the neglect by those running the government people in the area have developed survival skills and self-reliance. He commented that they don't need access to any formal institutions for this, they resolve community problems on their own. Here he was implicitly referring to a sort of informal justice system[4] used to deal with problems of thievery and cheating. He added that by their sheer numbers the government depends on his community for votes and must win and in this sense they have power collectively.

In contrast, Farouk, a very adept young Internet user and local community activist in Mamobi, used a similar reframing approach that enhanced the self-portrait he constructed of himself but at the expense of his community. Rather than defend his peers from vilification he stated plainly, "there are people who try to buy things . . . from outside like use credit cards to buy things . . . basically I would say majority of the people who goes to the café that is what they do. . . . Yeah, yeah, in our community I would say it's true . . . so I don't take part in all these things." By distancing himself from his peers in the community he claims a status of apartness and specialness that was coherent with the way he portrayed himself as a diamond in the rough, a young man with a disadvantageous background who was fighting the odds to make a success of himself.

Rumors define a boundary between insiders and outsiders by revealing that those who hear and believe a certain kind of rumor are members of a group of shared interest (Turner 1993). One rumor that was widespread among Internet café users in Mamobi was that I was not in fact a researcher as I claimed but rather a CIA agent. This rather abruptly positioned me as a hostile and duplicitous outsider from the perspective of the group. This rumor was another way in which young people reframed the way Mamobi was maligned as a zongo in order to establish their effectiveness in relation to the Internet. It built on the assertion that the Internet was a powerful way of making contact with foreigners. If CIA agents were showing up in Mamobi it meant that local youth had become very effective at using the Internet to establish contacts abroad, so much so that they were actually compelling foreigners to come to Ghana. It has been noted that rumors can have multiple interpretations and that individuals may tell the same rumor for different reasons (White 2000). The CIA agent rumor reflected not only a sense of efficacy, but it also reflected the understanding that Mamobi is seen by outsiders as lawless and criminal and that white foreigners and other outsiders who present themselves as allies may be adversaries in disguise. Although I heard this rumor in other areas of Accra, it was most pervasive in Mamobi where I was generating so much suspicion among the customers at one Internet café that the owner suggested that I should stop coming by so frequently and in fact I felt compelled to stop visiting that café altogether.

The case of the CIA agent rumor reflects how Internet users were preoccupied not only with issues of morality, but also with positioning themselves as effective in relation to the Internet. The efficacy of Internet use was constructed in rumors through casting strategies that amplified the perceived power of those connected in some way to Internet crime. Through the CIA agent rumor, I was recast as someone more powerful than was the case. Pushing characters to extremes this way had a dual benefit. It affirmed the efficacy of Internet crime by claiming contact made with the most powerful of foreigners, the largest sums of money, and even (by extension) powerful foreign government agencies such as the CIA. At the same time it maintained a sense of moral equilibrium because such powerful figures are impervious to harm. Through these stories embellished with hyperbole, the potential for power among local youth through the use of the Internet began to seem not only possible but also limitless. In this way stories about celebrities and other powerful characters as Internet crime victims are useful to establish the Internet user as simultaneously good and effective in relation to the Internet.

However, to be good and effective in relation to the Internet was a duality that was often difficult for Internet users to reconcile when they told stories about Internet crime perpetrators. As described in rumors, the most effective ways to make big gains on the Internet were often also challenging to justify on moral grounds. To get $20,000 from a foreign connection on the Internet likely meant doing something illegal and doing someone harm. Stories about Internet crime perpetrators forced the choice between good (through nonidentification) or effective (through identification). Internet users employed a variety of strategies in an attempt to reconcile this tension. Kwadjo, the nonscammer quoted previously, tried to bridge this gap by drawing a comparison between Ghanaian women legitimately finding foreign boyfriends online and Internet dating scams. He commented, "genuine people use the same procedure and it works," suggesting that rumors about criminal activities can also serve to affirm the efficacy of the Internet for legitimate uses.

Kwadjo's statement also points to an essential continuity between what young Ghanaians define as scams and what they define as legitimate Internet activities. In this way rumors about Internet crime could serve double duty, affirming legitimate and illegitimate Internet activities as effective. In addition to rumors about Internet crime, there were other rumors that circulated about online dating and Internet love. These took on the same structure as a foreign link leading to a big gain (in the form of a visa invitation) and resulting in a transformation through marriage and moving abroad to join a foreign spouse. For example Ahmed, an Internet café operator, described a friend who had met an American woman, chatted with her for four years and eventually moved to America to be with her. He noted, "I know one day I will get the person who will help me." And added, "Internet love, it happens." In all cases, rumors about big gains whether obtained legitimately or not, held to an assertion of efficacy stated explicitly in the comments, "it works" or "it happens."

Daniel is a good example of how the tension between good and effective Internet use could be contradictorily expressed by one person. When he noted that "in Ghana here we have some people who are having access to Mike Owens's account . . . ," he depicted himself as part of an imagined movement of young men in Ghana who are very effective with the Internet. However, moments later he revised his story adding "and I don't know who did that, but I learned it's from Nigeria not in Ghana . . . but not in Ghana, not in Ghana actually" relocating scams to a typical scapegoat group and thereby renouncing his membership suddenly midstory. Although admitting to attempting credit card fraud, he also claimed to have stopped and noted in a related story about the activities of his scammer friends that "I'm a Christian so I don't indulge myself in those things." Daniel was also the only person interviewed who used the term *rumor* as a label for his Internet crime story. By representing himself in a multiplicity of ways Daniel attempted to present himself simultaneously as an authoritative insider to Internet crime activities, a reformed former fraudster, as well as a savvy and skeptical outsider. His self-awareness reflects the complexity and ambivalence of being effective and good in relation to the Internet, something he seemed to tentatively accomplish by actively shifting back and forth in this single conversation between multiple perspectives.

The relationship between such tales and the actions carried out on the Internet can be further refined by considering the role such stories played in discontinuance decisions when an individual Internet user makes a decision to quit Internet use or adoption decisions when they first initiate use (Rogers 1995). For example, Daniel relayed several rumors in his interview but ultimately described how after attempting credit card fraud activities for a few months he quit, noting that "I involved myself in [it], but since I've seen that it's not beneficial to me and I have not get anything out of it, I stopped." For Daniel, the failure to realize gains from his investment of time and money (the accumulated evidence of direct experience) seemed to eventually tip the scales against the compelling stories of big gains communicated in rumor. Likewise, before ever venturing into the Internet café, Maureen had heard some rumors about the Internet café that dissuaded her from visiting. She noted that "before my first time of visiting the café I heard kind of when people go to the Net, kind of browsing pornographic stuff. . . . Yes, so it was like going to the café no, no, no I won't even try it." Yet, overcome by curiosity she eventually began visiting the Internet café regularly. Eventually she came to favor her own observations and experiences over these stories. She was insistent that "people who don't come to the café, they think people just come here for fun or I don't know just to watch naked ladies on the screen and stuff like that, but you see that that is not it at all." These reflections make visible the shifting equilibrium of Internet use and the possibility that either rumors or direct experience can eventually switch into or out of favor.

A few more things may be said about these rumors and, in particular, about their peculiar durability. The durability of rumors is demonstrated by their persistence over time and their capacity to move more or less intact across social space. Rumors were stories performed frequently and that circulated widely. There was something in the contents of these particular tales that, in the given social context, led them to become entrenched in memory and that compelled retelling. This made rumor something quite distinct from the more typically ephemeral and everyday speech acts. Among the many interpersonal spoken exchanges that take place on a daily basis, rumors endured a filtering and circulation process. As for the memorability of rumors, this surely stems from their depiction of the extreme, of what is most remarkable and out of the ordinary. The consequence was that cases of outrageously successful Internet crime (whether celebrated or lamented) were the sorts of stories told and retold and were therefore dominant in the public discourse. Experiences that run counter to circulating stories, such as Maureen's realization that the Internet cafés were not just for watching porn, did not feed back into message dissemination via rumor simply because the mundanity of such contrary experience made it a poor candidate for rumor mechanisms. In this way rumor circulations had a peculiar amplifying effect. One can surmise that although rumors in this way fed into the lofty aspirations of Internet users they also generated a disproportionate sense of what to expect as the likely outcome of their engagements with the Internet.

Orality in Contemporary Urban and Digital Domains

To further elaborate on speech acts and the Internet in Ghana this section briefly extends the analysis beyond rumor to consider how Internet users handled the potency of words (as a problem or an advantage) and acted on this in online environments. A broader context for such a preoccupation with words and their potency is observed in certain aspects of everyday interaction and at social events in Accra. In drawing from this context I wish to avoid altogether the clichés of speaking about the oral cultures of Africa but still emphasize the enduring importance of the art of speaking and of a local, particular, and long-standing appreciation of the "skillful control of words" (Yankah 1995, 45) even in contemporary, urban settings in Ghana. This was apparent in the many rich and varied forms of oral performance readily witnessed at church services, music or theater concerts, as well as woven into everyday conversation.

In an orthodox view of orality, scholars considered oral texts to be the reservoir of deep and long-standing tradition of a culture's self-understanding that stands apart from what is documented in the archives of colonial authorities, missionaries, and explorers but is also distinct from the transitory nature of popular culture (Finnegan 2007). Such works have often been portrayed as a distinctive offering of African societies in the diversity of world cultures and an area of "cultural achievement" (Finnegan 2007, 141). However, recent scholarship combating the ahistoricism underlying such an approach has sought to reposition oral genres as an evolving and contemporary phenomenon. The totalizing division between oral and written cultures has been challenged as overdrawn and inaccurate (Finnegan 2007; Barber 1997). Oral forms are equally a part of urban environs, are mediated by various technologies, and find their way into global circulations (Shipley 2009; Ebron 2002). The broader landscape of oral performance in Accra must further contend with new ambiguities such as text messaging and Internet chat where oral conventions appear in written formats and the two modes of communication collapse into one another.

In contemporary Accra, the question certainly does not come down to whether cultural production and expression are principally oral or written. The significance of oral forms must instead be situated among many other coexisting and, in some sense, competing formats. A written and visual mass media is firmly situated in Accra. Print publication has experienced a resurgence over the past two decades with the number of newspapers in circulation steadily multiplying at the newsstand. Informal publication of pamphlet literature and romantic and moralizing tales have also experienced a growth in popularity (de Bruijn 2008). There is a proliferation of advertising and other public signage, and abundant bureaucratic, professional, and legal documentation. Thus oral performance stands alongside diverse texts and literacies in the urban setting.

There are certain qualities, however, that Accra's urban denizens recognize and appreciate in the way oral genres in particular are performed. Witnessing such performances and the emerging audience commentary illuminates how speech is thought to function in social and material worlds. One illustrative example is the role of the master of ceremonies (M.C.), a figure who mediates social events including graduation ceremonies, outdoorings, funerals, traditional weddings, and so on. There are certain resonances between the role of the M.C. and the traditional figure of the *okyeame,* the royal spokesperson who speaks on behalf of the chief. Historically, the imposing aura of the chief is maintained by his literal silence (i.e., the chief does not speak out loud to the public), which is made possible by the use of a surrogate, his okyeame, to whom he delegates his public voice. The okyeame serves as a kind of diplomat enhancing the dictates of the chief through eloquent phrasing. Such eloquence is not appreciated simply on an aesthetic level. Speech is also understood to manifest as a force, one that the skilled are able to adeptly handle. The okyeame thus is seen as protector of the chief and his subjects from the potency of words (Yankah 1995, 10). The contemporary role of M.C. is likewise one of diplomacy because there are many pressures to be managed within such events, including family conflicts and intergenerational differences. However, in a modern twist, this skill has in recent years become commodified.[5] The M.C. is paid to work the funeral, wedding, graduation party, or other event and is one among many service industry workers at such social events including the DJ, caterer, videographer, and equipment supplier providing tents, chairs, and decorations. The value of the M.C., I was also told, was related to his capacity to persuade the audience to contribute to fundraising, a significant component of any social event that is meant to offset the costs of the ceremony and to contribute to those who are being celebrated as they embark on a new stage of life or, in the case of a funeral, to those left behind.

The M.C. must not only speak eloquently and with apt insight into the occasion, but is also admired for his unplanned speech and on-the-fly witticisms. I was the subject of one such witticism during the fundraising portion of an outdooring (baby-naming) ceremony. The M.C. stood exhorting the crowd to contribute. I had come dressed in a traditional kaba and slit in white, an outfit I had made up by a seamstress two years before for another event. The style I had selected was trimmed by the dressmaker with large fabric covered buttons. It seemed to me to be a bit plain by comparison to the thoroughly up-to-date and glamorous stylings I had seen on Ghanaian women more experienced with fabric selection and the dressmaking styles that were currently en vogue. Furthermore, at this event held at a home in the affluent East Legon neighborhood I found that a kaba and slit was not at all the going style, especially for young women of my age. Urbane young Ghanaians were instead dressed in more modern, casual white linen outfits. The mother of the newborn wore a button up shirt and trousers. As I stood to contribute my cedis to the collection, the M.C. presented me as "obruni from Kumasi" to the great amusement of the audience. It was a joke about my overly traditional attire (via reference to the city of Kumasi, the ancient capital of the Ashanti kingdom) juxtaposed humorously with my obvious foreignness as a blue-eyed, light-skinned woman. As this experience illustrates, the repertoire of the M.C. is not fixed, nor is it confined to the realm of local or traditional sociocultural relations, but it can accommodate the novel, strange, and foreign.

Such a tale from the field aptly captures the reforming and reframing of long threads of cultural continuity emerging in new ways in the urban field and with an awareness of being situated globally. Alongside such contemporary turns in traditional oratory, there are the new possibilities and challenges of online communication in which the oral and the written do not simply coexist side by side as they do in the urban space; rather, the two formats now merge together to become indistinguishable. In online chat, the instantaneous back and forth between co-present participants is akin to a live, verbal exchange. Yet the medium is typed words. There is also the materiality of chat software that lends to this orality of the written, via the instantaneity of message transfer and the rudimentary or nonexistent archiving (by comparison to email for example) that makes these exchanges appear as a linear and ephemeral dialog. The introduction of emoticons to represent facial expressions is a now well-entrenched user adaptation that emerged specifically as a phenomenon of Internet-based communication. It reflects another effort to make this format more like an oral form by reintroducing body language. In light of the blurring of oral and written there is a question of how established conventions or valued aspects of oral performance are called on or translated into this new space by Ghanaians who venture into the online world.

Among the different formats of Internet communication some variation follows from the materiality of these modes as well as the role played by administrators. In the particular case of online chat, there is an emphasis on expedience (through abbreviations) and informality. The format is one-to-one engagement rather than publication for a mass audience. It thereby occupies a realm of more casual, everyday speech. The moderated mailing list is quite another form. The first Internet mailing list on Ghanaian affairs was given the name *okyeame* making explicit reference to the refined and eloquent speech of the royal spokesperson and potentially marking the list as a space for more formal communication. Administrators of the list note it was named as such "because the network serves as a medium through which different people communicate—a functionality that is similar to the way chiefs and people communicate in Ghana."[6] In other words, the mechanism of the list itself was the okyeame, the diplomat mediating between different members. Furthermore, the moderators of the list also take on part of the okyeame role of diplomat and mediator in the socio-technical system by regulating conflict, encouraging productive and on-topic discussion even if it sometimes meant banning unruly participants.[7] This particular mailing list is a space of communication generally among highly educated Ghanaians in the diaspora and the list itself is archived and distributed from servers at the Massachusetts Institute of Technology, where several list cofounders attended as graduate students. Those who wish to join must be vetted by list administrators. All of this runs against a particular ethos of online environments (once dominant but now seemingly diminishing) favoring the unencumbered prerogative of the individual and totally uncensored self-expression (Turner 2006; Castells 2001). Heavy moderation and efforts to circumvent anonymous communication are much more in keeping with a recognition of the potency of words and the need for communication to be carefully facilitated in order to smooth and maintain relationships. In this way the adoption of the mailing list by these Ghanaian elites who came together in explicit recognition of a shared heritage incorporated in its design and management a certain sensitivity to this potency.

A belief in the potency of words extends also to the way some young and nonelite Internet users in Accra thought about the efficacy of their digital interactions. This was demonstrated in the comments of two young men. Both were once regular Internet café users though had lately become somewhat disengaged as the offline demands of job responsibilities impinged on nonessential uses of their time. They conjectured about what was necessary to use the Internet effectively and attempted to account for their own failure to generate outcomes, particularly with scamming strategies. As Daniel noted of scamming, "it works for some other people because they're having a *sugar coated words* and I'm not having it . . . they have words that convince the whites" [emphasis added]. Daniel's friend Kwaku in a separate conversation also noted "sometimes they use *sweet words* . . . they can get the site, lovingyou.com. So some of them can use loving words to tell the person they love you. . . . The person will fall in love with you . . . they don't use any magic, but it is the words that you use because *words are powerful*" [emphasis added]. He noted that his fellow schoolboys would look for words that are "romantic and powerful" on such love quote Web sites. His example was this quote professing love and singular devotion in poetic language: "every girl caught my eye, but you are the only girl who caught my heart." The measure of success was monetary. If one was successful with such words, eventually the seduced foreign target could be told, "'I need this amount of money,' or maybe, 'I'm in financial problem,'" with the outcome that "the person will send you money without even you telling the person the amount you want." These comments reference another process of commodified speech performance, similar to the urban M.C., though in this instance carried over into the digital domain. An instrumentalist sensibility is demonstrated in the view that words properly deployed can generate tangible outcomes. These young men position persuasive speech as a form of technical work on the sociotechnical system of the Internet. Whereas the genre of rumor reflected how "social truths" circulating external to the Internet could have material consequences for the way the Internet was engaged, these examples show how words also came to be consciously and directly deployed by Ghanaian Internet users within online environments.

Conclusion

When young Ghanaian Internet users sat before the screen at a local Internet café, they brought with them a range of resources for making sense of the machine, including a history of their own experiences with other mediation technologies (such as foreign pen pals contacted via postal mail), their prior experiences with the Internet, and also the real and imagined experiences of their fellow Internet users in Accra. Rumor was a mechanism that propagated these real and imagined experiences of others yielding a persistent social imaginary that resolved questions of the efficacy and morality of Internet use practices.

Through the analysis of rumors this chapter has considered some larger issues of how language is materialized, how it registers a consequential force in speech performances, and to what ends. Speech relies on the acoustic capabilities of the human body operating in concert with human memory. Within the range of speech genres, certain formats such as rumor are especially durable, worming their way into memory and compelling those who receive the stories to retell them. They are viruslike, adaptive and opportunistic. Rumor as an oral genre contradicts the way spoken words in material-semiotic analysis are often relegated to the merely corporeal and ephemeral. The potency of words, their ability to supplant direct experience in this case, challenged the notion that physically manipulating technology was necessarily weightier, more material, and thus more believable or more consequential than speaking about technology. In conjunction with material manipulations at the computer interface, words materialized in human speech may fill the gap between experience and expectation.

The examination of rumors about the Internet has also accounted for the way users may influence one another toward an emerging, collective understanding of the technology. Rumors were not merely idle talk, but were stories that were believed in, and that offered a general structure to the actions carried out at the machine interface. The compounding and amplifying effects rumor had on its contents gave it an autonomy beyond any individual. This capacity users have to develop distinctive shared interpretations and unanticipated uses of a technology has been well accounted for (see Fischer 1992; Kline and Pinch 1996; Bijker 1995). However, analysis in this mode has generally not considered the mechanisms that propagate these interpretations among members rather pointing ambiguously and implicitly to shared socialization or common circumstances as their source. This chapter proposes alternately that users play an active role in shaping one another's views through accessible communicative mediation. The distinctive materiality of such mechanisms as rumor also shape these interpretations. The amplifying effect of rumor yielded the reproduction and overrepresentation especially of those dramatic and memorable stories of big gains and thus reinforced a persistent, go-for-broke approach to building a social network online and the enrollment of foreign contacts among young Ghanaian users.

One purpose of reviewing rumor, speech acts, and orality in its various forms and instantiations in the urban context of Accra, Ghana, is to show what sorts of alternate epistemologies come to the fore once we move into regions that have not been well incorporated into the existing models of technology's uptake and adaptation. The confidence placed in the social truths of rumor relate in part to the marginality of this user population who turn not to authoritative sources (in the context of a mistrust of or neglect by authorities) but to ones that are more relevant and close at hand. An explicit acknowledgment of the potency of words among urban Ghanaians has brought certain processes of sense-making forward that have been overlooked in the circumscriptions of more mainstream work in STS. A similar approach is taken in chapter 5, which looks at another domain in the forms of religious practice that range from attending church and observations of prayer to the religious syncretism in services rendered by a Muslim mallam who draws from an indigenous animism while incorporating elements of the Koran. In these spaces the Internet and its functionality were also engaged and explained through practices of second-order sense-making apart from the immediacy of the machine interface. This domain further complicates notions of materiality by introducing supernatural force as necessary to account for in local understanding of the efficacy of Internet practices.

5 Practical Metaphysics and the Efficacy of the Internet

Religious practice and belief were a frequent point of reference for Ghanaian Internet users when they spoke about their social relationships, aspirations, and their use of technologies including the Internet. The way they talked about this belief was marked by a sense of the presence of spiritual forces (good and evil) and the operation of these forces in one's day-to-day existence. This chapter delves into the complicated and dynamic terrain of religious practice in urban Ghana. To refer to this as a matter of "religion," however, is in certain ways misleading. Sociological approaches to the study of religion since Durkheim (1995 [1912]) have generally focused on religious institutions as a form of social organizing. The central concern of this chapter is rather with a foundational metaphysics, a particular shared body of knowledge about how natural, manufactured, and supernatural forces operate in the world. This cuts across various religious affiliations in Ghana but often finds its concrete form in particular religious practices and in spaces of religious ritual.

Chapter 4 was concerned with how issues of the morality of Internet use were managed in the construction of a social imaginary in rumor. The question of morality, perhaps surprisingly, was *not* how the religious practices of Internet users functioned in relation to the technology in Ghana.[1] Rather the concern among users was principally with efficacy and how this was tied to the management of spiritual forces. The complex interplay of supernatural, material, and specifically technological forces (including those of the digital realm) was a de facto reality that many Internet users attempted to intervene in and to delegate between. This meant that what sometimes appeared to be rigid persistence in certain, so far, ineffective ways of engaging with the technology (such as in strategies to seek financial gain or opportunity through online contacts) were explained by adjustments that users were actively making in other parts of the system. This included efforts to undertake rituals of spiritual alignment and enrollment by visiting a fetish priest or mallam or by attending church services. As with chapter 4, this analysis builds on the larger point about the incompleteness of an analysis circumscribed too closely around the machine interface. In this case I argue that to understand how the Internet was distinctively materialized in Accra required going to church.[2]

In a material-semiotic approach, how the supernatural realm may be accommodated analytically is largely uncharted terrain. Prior case studies of actor-network theory have generally been set in a disenchanted West and often at sites of technoscientific production, where there are certain rigid a priori commitments to empiricism and to certain prevailing notions of what is real and unreal. However, a belief in supernatural force as it is translated into observable material consequences is clearly under the umbrella of the "practical metaphysics" of actors that Latour indicates is the concern of a properly materialist approach (Latour 2005, 50). As a pragmatic methodological matter, to pursue the supernatural in this way is to trace out what users consider to be its material evidence, the signs of supernatural intervention in this world, and consequent actions that are undertaken in response. The question of whether such forces really, truly exist or are illusory is not the concern of this analysis. To those living within this metaphysics, I found, the question was rarely one of whether such forces exist, but of how to distinguish good and evil forces, the trade-offs between them, how to enroll these forces, and whether or not it was necessary toward one's aims.[3]

There is a particular significance in the matter-of-fact way that religion and the supernatural were accommodated by users in engaging with the Internet in Accra in relation to the book's larger concern with scholarly work on the global transformations of the digital age. In particular, what was unfolding in Accra refutes one line of thinking that suggests network technologies have become a secularizing force among the populations that embrace them. A related discussion considers processes of democratization to be (under the right conditions) the expected effect when such capabilities are introduced in regions of the Global South (e.g., Ott 2001; Ott and Rosser 2000). On the issue of democratization, researchers who study the Internet as a space of political engagement have frequently turned to

Habermas's definition of the public sphere as a heuristic device. Habermas defines the public sphere as a space open to all comers where participants relate as equals engaging in rational debate apart from the instrumental demands of business and economic survival and free from interference by state authority (Habermas 1991 [1962]). Researchers have used this ideal to analyze whether the rapid diffusion of the Internet marks the re-emergence of the public sphere out of an era of political apathy and corporate-controlled media in the Global North and potentially out of the repression of political speech by the state in the Global South (Rheingold 1993; Herring 1993; Poster 1995a; Ess 1996; Eickelman and Salvatore 2002; Bernal 2005). Habermas addresses religion directly in his argument indicating that in the wake of a functioning public sphere religion will increasingly become marginalized and privatized because ostensibly it is incompatible with a purely rational mode of discussion. He judges that this process of secularization is a desirable continuation of the Western Enlightenment project. In accordance with his thesis, to the extent that the Internet functions as a public sphere and becomes more and more central to the public life of a citizenry, we would expect religion to retreat to private and dedicated spaces—the home as well as churches, mosques, temples, and shrines.

Habermas's secularization thesis in particular has been compellingly and convincingly refuted in recent years (Casanova 1994; Asad 2003). Anthropologists have widely observed the highly Eurocentric nature of his argument and its inapplicability in other geographies (Hackett 2005; Meyer and Moors 2006). Habermas's secularization thesis marks a (by now) classic modernist argument opposing religion to a universal rationality. The basic structure of this opposition also appears in more recent scholarly work on the global transitions of the digital era. A notable example is Manuel Castells's examination of religious fundamentalist movements. Castells describes these collectives of Muslims and Christians as attempts at "constructing social and personal identity on the basis of images of the past and projecting them into a utopian future, to overcome unbearable present times" (Castells 1997, 25). He depicts a process of disconnection, from secular society and from global processes. These movements, he argues, constitute a reaction to and willful rejection of globalization, a process fundamentally facilitated by new network technologies such as the Internet (Castells 1997). His argument places religious belief and practice in opposition to new technologies, analogous to the opposition between tradition and progress, the past and the future.

Castells's formulation, similar to Habermas's, offers a notion of religion that I argue is excessively sociological. For Castells, religion is social affiliation, a form of membership. For Habermas religion appears as a remarkably fragile set of ill-formed understandings of the world taken on faith that are maintained only perhaps by the insularity of its believers. This misses an important aspect of how religious belief was pursued in Ghana, where religion encompassed a deeply felt and observed set of interconnections between visible and invisible forces, between the physical and spiritual realms. Practices to aid believers in operating on and in the world accompanied this underlying metaphysics. Religion, in a way similar to the Internet, was viewed as a system that individuals could operate to realize certain desired outcomes. Anthropologists have previously recognized that religion frequently functions as a technology, a theory of forces that can be applied to social relations, the material world, and the self (Gell 1988; Malinowski 1978 [1935]; Miller and Slater 2000). This is a partial explanation for what Internet users in Accra recognized as a compatibility between religion and technology. It is the theme of religion as technology (not religion as identity or religion as sociopolitical strategy) that resonated most with the many diverse forms of religion in Accra. A close study of two cases in particular (one a Christian pastor, the other a young man relying on the syncretic services of a mallam) will highlight some of the diverse configurations users constructed that welded technology to religious practice.

A Brief History of Religious Movements in Ghana

Given the complex syncretism of imported Muslim and Christian beliefs with long-standing indigenous beliefs in Ghana and in light of the numerous and increasing number of Christian sects that have established a presence and following in the country, it is necessary to devote a few pages to untangling the main threads and tensions in this religious landscape. Religion is intertwined with the region's political history and its shifting ethnic and geographical ordering. When speaking about their religious affiliations Internet users often described Ghana as a setting of religious diversity and tolerance. Indeed, Ghana has not been marked by the episodes of violence in the name of religion that have afflicted neighboring Nigeria. At the same time, there was competition between religious groups with a huge variety of Christian denominations represented in Accra. In the course of this research, when asked about their religion, most Ghanaians identified themselves as Christians consistent with the most recently available statistical data that suggests a predominantly Christian population particularly in urban areas (Meyer 2006). In rural areas an indigenous animism is more widespread and is practiced more openly. The northern part of the country is home to a large Muslim population. Among the Christian denominations present in Ghana, researchers have noted a primary division between "mainline" or orthodox churches (Presbyterian, Methodist, and Catholic churches among others) and the more recent rising popularity of Pentecostal, charismatic sects. Pentecostalism emerged through the 1990s as the main current in Ghanaian Christianity and by 2000 a good 46 percent of inhabitants of the greater Accra area claimed to belong to Pentecostal denominations (Meyer 2006). The neighborhoods of Accra that housed significant Muslim populations (such as Nima, Mamobi, and Newtown) were among the poorest in the city and Muslims constituted an often maligned religious and ethnic minority. Although religious orientations showed geographical tendencies in Ghana,[4] they were not strictly bounded by geography any more than the citizens themselves, and particularly in the urban capital people with various religious affiliations intermingled on a daily basis.

Islamic practices in conjunction with the Internet were not as readily apparent as Christian practices in Accra and therefore are not covered extensively in this chapter. Much of the research for this book was conducted in Mamobi and many Muslims were interviewed in this site. Religious rituals such as Friday prayers and the pilgrimage to Mecca were discussed in these interviews but in a way completely detached from uses of the Internet café. The lack of material relates, in part, to the suppression of these religious followers by the evangelical Christian majority. Muslims made it clear that there were widespread negative connotations to their religious affiliation in Accra. They were less eager to speak about their faith in interviews and did not have the same imperative to proselytize and seek converts as many Christians. The lack of material is itself an indication of the place Muslims have in Ghanaian society, one that is marginalized and kept private but is not overtly under attack by the government or citizenry.

Anthropological research has highlighted tensions in the religious landscape in Ghana as competing practices and beliefs evolve (through individual and collective efforts) to address contemporary concerns. Three issues in particular were actively contested in churches, the mass media, and other public spaces in Ghana. The first was a tension between the way emerging strains of religious belief critique and resist but also facilitate processes of modernization and globalization. Second, although many Ghanaians consumed diverse religious media (music, sermons, movies) and expressed an openness to alternate religious practices, there was also a vociferous public debate about competing practices (Christianity versus indigenous animist strains of belief, for example) and whether they could be effectively and legitimately mixed. A third tension was between the focus on salvation versus prosperity. The first Protestant and Catholic missionary churches in Ghana emphasized modest living and the promise of rewards in the afterlife. The growing popularity of American Pentecostalism presented a challenge to this perspective touting the prosperity gospel that links religious faith to rewards (spiritual, social, and material) in this lifetime (Gifford 1990; Hackett 1995; Meyer 1998a). Far from serving simply as the domain of unchanging tradition, the scope of religious practice was marked by the powerful undertow of trends that were progressive, pragmatic, and syncretic adapting to and reframing ongoing social and economic changes in Ghana.

The public role of religion in Ghana was closely tied to popular culture. Religious content (mostly Christian) was pervasive on the radio and television and in video films, CDs, and cassette tapes. Religious practices were also highly visible and audible in the streets of Accra. This was primarily true of Christian churches that commonly amplified their sermons and music on Sundays as well as on weeknights to capture the attention of anyone nearby.[5] Local music forms frequently conveyed religious messages and gospel music videos were shown daily on network television. Ghanaian and Nigerian video films, very popular in Accra, dealt primarily with the religio-supernatural cosmologies that fascinated many Ghanaians. These videos were often heavy on spectacle (particularly the Nigerian ones) and typically dealt with the machinations of evil forces. Videos supplied information about the workings of the supernatural realm and inspirational or cautionary tales about the triumph of good over evil and the proper moral path to ensure safety and genuine prosperity (Meyer 2006). Videos as well as popular literature were widely consumed for their entertainment value and also their educative role on matters of religion and morality (Newell 2000).

The prominence of religious content in the mass media is evidence that modern technologies have not simply compromised the role of religious faith in Ghana but rather provided new platforms for its diffusion. Evolving religious practices have also addressed an increasing reliance on and desire for foreign material goods including electronics, cars, clothing, and household items. Meyer has examined the perception among Pentecostal Christians that these foreign imports present a threat. The danger is that if one is not vigilant one's possessions can potentially possess the owner in the most tangible sense: cars can take control from their drivers, underwear can cause unwanted sexual dreams (Meyer 1998a; Verrips and Meyer 2001).[6] Through religious practices among Ghanaian Pentecostals, such as prayer over "enchanted" foreign consumer goods, the threat they pose is eliminated facilitating their unencumbered use (Meyer 1998a). Parish describes the acknowledgment of this danger via religious practices as a reflexive consumption that engages with the ambivalence of change and of a constantly expanding and diversifying material culture whose origins are increasingly obscured, a distinctive twist on Marx's notion of commodity fetishism (Parish 2002). Religious affiliation, with its accompanying code of ethical social conduct, also eased fears about the new possibilities of social ties to strangers through new communication technologies like the Internet. For some Internet users in Accra, religious faith and trust in the fellow Christians they encountered online yielded a sense of protection and safety in virtual space, that is, that they were being watched over by a divine presence and would not come to harm.

In their competition for members, churches and other religious institutions were under pressure to recognize and address contemporary and worldly concerns. Christians in Accra described something akin to a religious "consumer culture" with churchgoers discussing often subtle preferences and understandings of differing styles of worship, systems linking faithfulness to goal attainment, the appeal of certain charismatic preachers, and pragmatic church programs that promise assistance with job searching and education. The urban setting provided good access to radio, television, and other communication technologies as well as a great variety of churches and other religious institutions. Many of the individuals I encountered during the course of this research made use of these diverse resources to sample and compare religious teachings. Many had already converted from the church of their childhood to another of their choosing. Conversion from Christianity to Islam and vice versa was not unheard of, although it was more rare. Pastors, shrine priests, mallams, and others who provided religious services correspondingly recognized the need to appeal to potential clients. Parish described competing advertisements among shrine priests of the particular gods they represent and the unique skills these gods hold and assistance they are able to provide to those who visit their shrines (Parish 2003). Traditional figures such as these shrine priests competed alongside the more recently arrived world religions to attract clients.

The diverse religious influences in Ghana and the competition among various denominations created many opportunities to combine the practices of competing religions. There was a tension around whether this was legitimate and moral or whether it was imperative to maintain purity of practice. Mixing indigenous animist and Christian practices was particularly controversial. Pentecostal faiths emphasized the birth of a new self that involved the full rejection of these animist practices (such as the use of talismans and fetish) as the work of the devil (Meyer 2006; Meyer 1998b). However, although pure Christian faith was publicly proclaimed, an underground market of treatments by fetish priests and through shrine rituals remained active. The pious Christian who secretly seeks the assistance of such fetish priests for problems with illness or infertility (portrayed as dangerous, dishonest, or even humorous behavior) is a locally familiar archetype, one frequently portrayed in Ghanaian and Nigerian films.

Furthermore, faithful Christians and Muslims did not always recognize that they were transgressing the teachings of a particular faith because old and new beliefs have, in some ways, merged together and become indistinct. Christian denominations that have gained popularity in Ghana have undergone a process of adaptation where in various ways they have been made coherent with a pre-existing ordering of natural and supernatural worlds. The resulting religious syncretism of this adaptive process is common to the emergence of world religions throughout Africa (Horton 1971). Therefore despite the exhortation to reject old ways, what took place in Pentecostal Christian churches (and to a lesser extent the mainline Protestant churches) in Ghana was a syncretic blend of new beliefs with pre-existing worldviews. For example, rather than reject witchcraft and fetish practices as figments of the imagination, these beliefs became incorporated into a Christian worldview as the work of the devil (Meyer 1995). The harmful activities of family members against one another (a key theme in witchcraft narratives in Ghana) and the effective (albeit demonic) use of fetish often featured in stories told by pastors in their sermons and by ordinary church members in their testimonials. Even for pious Christians in Ghana, witchcraft and fetish often remained very real, very dangerous forces, and therefore still played a role in their everyday lives.

The way a Pentecostal metaphysics seems especially able to accommodate aspects of a prior indigenous belief is offered as an explanation for their rapid and growing popularity in Ghana (Meyer 1998b). Pentecostalism was positioned by followers in Ghana as the most effective tool for battling supernatural adversity in well-known, long-standing forms. Furthermore, these emerging Pentecostal churches were appealing because "they claim to have the answers to Ghanaians' existential problems and especially to their most pressing existential problem, economic survival" (Gifford 2004, ix). Pentecostal churches promoted a distinctly this-worldly approach that was not limited to securing a spot in the afterlife but was also concerned with achieving material success while still on earth. A tension between the focus on salvation in the afterlife versus prosperity in this life emerged from competing Christian worldviews. Notably, because the prosperity gospel emphasized rewards realized in this lifetime, it could be tested because the consequences of religious practice could be observed before death. It was also coherent with an animist worldview that has always been concerned with ways of operating (via the supernatural) on the natural and social world to realize desired outcomes. The way individuals adopted, combined, and abandoned those religious practices that proved ineffective was, in part, a process of testing. The testability of a this-worldly orientation contributed to the way religion was employed as a technology in Accra.

The Internet and Technology in Church Sermons and Testimonials

Continuing with themes from the previous chapter affirming social belief and the significance of spoken genres and powerful words in urban Ghana, the church sermon as well as church member testimonials were additional forms of small media similar to the rumors examined previously that played a role in making sense of technology. In rumor as well as church communications, messages were relayed verbally and in person. Both set expectations and potentially shaped behavior relying on what were thought of as true stories for their impact. However, the communications in the church diverged from rumor in certain ways. Believability was communicated, in part, through the hierarchical mode of the church sermon in which the message is sustained by the preacher's authority and charisma. The relationship was one-to-many rather than one-to-one as in the case of rumors.

The experience of attending a church service was a richly multisensory and embodied experience. It involved the use variously of anointing oils, sacred books, of choreographed call-and-response routines, and other techniques for transferring power between preacher and congregation. Particularly in the newer Pentecostal movement there was an emphasis on tools for battling evil, shifting at times into a kind of how-to seminar. For example, at the large megachurch Royalhouse Chapel led by Reverend Sam Korankye Ankrah, in the midst of a testimonial about a reformed thief, the preacher turned to the congregation to inform them that "here at Royalhouse Chapel we teach people how to pray. This is the way we do it." He instructed that if the devil came into your marriage, your home, into your business or finances, to "clap your hands and say Out! Out!" The congregation dutifully responded cheering, "Out! Out!" and rapidly clapping. He followed with this:

Reverend: Clap your hands and say "father."

Congregation: Father! (collectively and rapidly clapping now)

Reverend: In the name of Jesus . . .

Congregation: In the name of Jesus . . .

Reverend: What is formed against me . . .

Congregation: What is formed against me . . .

Reverend: Fashioned against me . . .

Congregation: Fashioned against me . . .

Reverend: Designed against me . . .

Congregation: Designed against me . . .

Reverend: In my house . . .

Congregation: In my house . . .

Reverend: Let that one fail . . .

Congregation: Let that one fail . . .

Reverend: In the name of Jesus!

Congregation: In the name of Jesus!

He concluded by noting, "Can you imagine if we pray this simple prayer every morning? The Bible says we don't receive anything because we don't know how to pray!" In another service the reverend physically mimed throwing such spiritual force, as though pitching a baseball, for the congregation to reach up their hands to receive. Such embodied prayer makes clearer a sense of how religion and its practices in this context are a manipulation of supernatural forces and a way of grappling with one's efficacy in the world. This underlines once again the incompleteness of treating religion as simply social affiliation (a club to belong to) or moral encoding (a set of rules taken on faith). The clapping and other bodily practices in these church services, speaking in tongues (common in Pentecostal churches), and cases in which afflicted persons are touched in order to be cured or anointed with oils by the preacher—all of this adds additional dimensions of materiality to such practices that were allied to the force of speech in sermons, prayers, and testimonials.

Technologies (old and new) occasionally were referred to in sermons and testimonials. The particular message conveyed depended on the church and whether it aligned with this-worldly or other-worldly orientations. For example, at a Presbyterian church in Accra, the sermon one week centered on the theme "God's unchanging word in a changing world." The pastor grouped new technologies including the Internet together with obscene music and the World Trade Center bombings as threatening changes that Christians could seek respite from in God's word. This was coherent with the tendency in the mainline Protestant churches to emphasize an alternate spiritual realm as distinct from a *worldly existence* (a term used pejoratively) though seemed strikingly out of step with the yearnings apparent in urban youth culture in Accra. By contrast, at a Ghanaian charismatic church I attended in the outskirts of London, the pastor, addressing a well-educated and cosmopolitan audience of Ghanaian immigrants, exhorted church members to embrace and learn about new computing technologies. He specifically called out Microsoft PowerPoint and Word in his pragmatic message about the work skills necessary for prospering in the West. At this church a laptop and projector were set up to show song lyrics and church events as a demonstration of the new way the church was advocating (Burrell and Anderson 2008). Large Pentecostal churches I visited in Accra were even more elaborately outfitted for video projection, recording, and amplification technologies as part of the infrastructure for Sunday services.

A testimonial offered at the Royalhouse Chapel in Accra provided a similar account of maintaining a hold in the technological domain. The church member stood up to describe a dream he had about the electronics he kept in his room being stolen. He interpreted the dream as a prophetic warning and decided to move his belongings. When he told his landlady about it, she was unworried. One night when he was at a prayer service at the church, someone broke into the house and stole many of the landlady's belongings, but none of his because he had heeded his prophetic dream and moved the items. The landlady, he noted, belonged to a competing church so his testimonial communicated a message about the ineffectiveness of competing faiths and their inability specifically to aid technology consumption. The young man treated the technologies he owned as prized objects (setting them apart from other types of belongings) and his religious faith had a direct impact on his ability to maintain this desirable state of ownership. His dream and the actions he took in response also reflect how the ownership of such goods generated a vulnerability that required extraordinary measures, even supernatural intervention to cope with.

These examples point to technologies in general and computing technologies in particular as encompassed by and subject to this larger system, this metaphysics of force. The appeal of world religions in Ghana was also shown by the international networks they tied members into. Another reading on speech and its power is taken up in the next section, specifically in the way the universal language of Christian faith in particular offers a kind of code for social bridging, a code that could be deployed online. This emerged as an alternative strategy that can be compared to the more cynical persuasive speech of Internet scammers as considered in chapter 3.

Networking Christians and Christendom as a Network

Most mornings before heading into the hot and dusty traffic jam between the home where I was staying in East Legon and the city of Accra itself, I would gather to pray alongside the family of Arthur, the born-again Christian pastor who was my host. His prayer, addressed to Jesus, included the ordinary appeals for protection but made one additional request. The pastor asked Jesus to "put us into contact with whomever can help us." Much like Internet use, religious practices were perceived as a set of techniques allowing individuals to extend themselves exerting a more powerful force of influence over broader social terrain. Both served in the expansion of agency. In this sense, if one fulfilled all the practices to gain the favor of Jesus (by observing Christian norms of conduct in relation to family and friends, attending church, tithing personal income, etc.) then Jesus would, in turn, become the ultimate patron, an all-knowing, all-powerful figure networking behind the scenes on one's behalf.

Connections between the social and supernatural (and back again) were possible through proper Christian practice, but even without supernatural intervention Christianity offered a language and a social code that improved one's chances for connecting across cultural and geographic distances. Christianity and the Internet served to expand the human social networks of those who appropriated these systems. The realm of Christendom is characterized by a common language, or in Castells's conceptualization, a mutually understood communication code (Castells 1996). A visit to any church service of any Christian denomination in Accra will illuminate these codes in the expression of Christian aphorisms such as "Jesus saves," "with God all things are possible." Christianity supplied a way of speaking with coherence and recognition across cultural boundaries through the common ground of Christian faith, something that Internet users found useful online.

Alexander, a pastor who frequently visited the La Paz Internet café, explicitly recognized this formation of Christendom as a network. He was among the few I encountered in Accra who had produced his own Web site. On the main page of the Web site Alexander and his wife are pictured in a wedding photo along with some basic information about the church including their weekly schedule. The site was designed not only to advertise the church he had started, but also to offer guidance and inspiration through selections from the Bible and pastoral commentary and advice. Another page, aptly titled "Let's Network," begins with the story of the Good Samaritan. The text continues with an appeal to Christians of all denominations dotted with references to "common ground," agreement, and fellowship,

> Friend, God gave us Christendom a common ground on which all of us can trod safely, and that is believing in Jesus Christ the only Son of God, through whom salvation has come to the world.
>
> The above statement is sufficient for you and I to agree upon no matter our affiliations, so that we can all have fellowship with one another. In that case, we would keep you posted by sending you our prayer request from time to time. You can also be a Financial Supporter to this work.

The pastor's approach employs a sense of sameness through religious affiliation presented as overriding any other memberships. He makes use of the special language indicative of this membership to appeal to site visitors' sense of belonging. Through the Good Samaritan story he reminds his fellow Christians of their obligation to assist. He segues into requests for financial support. Alexander followed this text with a list of desired technologies for the church including electronic music equipment and a computer, printer, and scanner as well as plots of land for building a church auditorium, chairs, and help with finding jobs for church members.

Despite the focus on the material needs of the church, Alexander envisioned what he was inspired to do on the Internet as an exchange. In his role as pastor he had something of value to contribute in his online encounters. He said that he spent most of his time online giving support and praying with the foreigners that he met in chat rooms. He described in detail several relationships he had established online talking with people about problems in their romantic relationships or with family and those suffering from illness. In ongoing conversations with an American man, Alexander had advised him on his relationship with a woman from the Philippines, her unexpected pregnancy by another man, and an engagement that was accepted and then broken. He described himself as an educator and counselor online, commenting, "I'm trying to project that idea to the Internet, to people that look if you learn to walk in the presence of God with God you will not, no matter what you are going through, you will not feel rejected, every day you will be experiencing the presence of God." Although aligned with the "prosperity gospel," his comments expressed a side of his religious faith that rejected the necessity of material gain for spiritual well-being. Furthermore, through his role as counselor he positioned himself in a much more egalitarian relationship with the foreigners he encountered than the relationships described by many other Internet users in Accra that were often oriented primarily by patron-client and dependency relations. Through his Christian affiliation, particularly his status as a pastor, he was able to subvert the social hierarchies defined by his nationality, age, and economic status that might otherwise come to define his online interactions.

Alexander described the benefit of the Internet in collaboration with his Christian faith. Neither the Internet nor Christianity operated as effectively without the other. Both facilitated strong, sustained, social connections but in different ways. The role for technology was one of amplification, a way of broadcasting his message more widely to reach more people. He noted that "I can use the Internet to reach the world . . . to sell out the ideas to people across the world that, look this is what the lord is using us to do . . . these are my visions." The Internet was not the only technology used to amplify his message; he also sought support for purchasing (literally) amplifiers. A practice common in many Pentecostal churches in Accra was to use speakers to blast Sunday sermons so that "the word of God" would reach those nearby who were not yet believers. Alexander's church service was, in fact, the loudest I ever attended while in Ghana amplified at ear-shattering decibel levels on low-quality speakers that desperately needed to be replaced. He described his desire to "hold the crusades and in Ghana here when we say crusades we mean mass evangelism. Right, you need powerful instruments, powerful machines that amplifies, which we don't have. Powerful speakers which we don't have to mount them up and preach the truth of Christ. And that is my vision . . . using the media to touch the world." The Internet was placed on a continuum with stereo equipment—speakers and amplifiers—as a technology for carrying messages across time and space to reach as many people as possible.

With technology in place for amplifying his message, Alexander's Christian faith played a role in improving the quality of those new connections by relying on the obligation to be connected in a way that was inherent to his Christian ideology. To approach the Internet as a Christian seeking other Christians was seen as a way to resolve the troubles many Internet users faced in finding a foreign chat partner who was *serious*. This term was applied to characterize a person committed to an ongoing relationship that potentially included tangible forms of support. Even Christians in Accra who had not made the church their livelihood pursued similar strategies, using their faith as a filter for Internet content that would steer them toward a more effective, less dangerous, and higher-quality communion with foreigners. Several emphasized that they spent time in Christian chat rooms as a way to find compatible chat partners and to avoid the immoral temptations of the Internet.

In describing his efforts to participate in Christian fellowship online Alexander underlined the role the Internet played in bridging vast distances. Christianity had the separate role of eliminating difference by serving as a common bond that superseded all other affiliations and identity markers. As he commented, "The trust is there [on the Internet] to accept one, no matter the race, no matter the tribe, no matter the nationality, no matter the continent. You know, it brings you together, it bridges you, you know . . . it makes you a family and at the end of the day you know that you're not alone even if you don't have anybody else to talk to, you can talk to somebody, Christian brother, far away, in Africa or in America. Or somewhere else in the part of the world." His description evoked classic utopian idealizations of cyberspace. He portrayed the Internet as a tool that allowed for the transcendence of an identity undesirably marked by race and nationality. This viewpoint echoed a hopeful and utopian view of new computing technologies also seen in IT marketing messages (Nakamura 2002) and in some of the early studies of the Internet (Rheingold 1993; Turkle 1995). This vision was not common among the Internet users I interviewed and (demonstrating the range of viewpoints) was a direct contradiction of the approach taken by cynical Internet scammers who saw the Internet as a space for manipulating foreigners' prejudice and lack of understanding. Alexander, by contrast, placed his faith in the capacity for human relations to be transformed for the better but not by the Internet alone. Similar to the Internet scammers, he recognized technology's inability to single-handedly correct the indifference of the Western world and instead saw Christian ideology as the solution. Alexander's vision joined the Internet and Christianity as two complementary components of a larger project of social connectedness.

Although Alexander principally described the advantages of his Christian faith when using the Internet in social terms, he was also explicit about its suprahuman dimensions and the link between Christianity and the Internet was clearly more to him than a sociopolitical strategy. Newer strains of Pentecostal Christianity and indigenous beliefs predating the arrival of the world religions rely on a vast ontology composed of entities and forces that are ordinarily invisible to human perception and that can be accessed and influenced only through special mediation. For faithful Pentecostal Christians an elaborate demonology that draws from indigenous animist beliefs makes up a significant part of this ontology (Meyer 1998a). Technologies and other material commodities are also generally understood as embedded within this ontology and are subject to supernatural forces. Meyer provides an example in her examination of the video production industry in Ghana where she notes concern over videography equipment being tampered with by evil forces in the course of filming stories with a strong religious message (Meyer 2006). A few Internet users who I interviewed believed these sorts of supernatural forces could be manipulated with beneficial effects and discussed the possibility of using them to make their online interactions more effective. The perception of this spiritual realm and how its forces relate to technology was an important aspect of the connection drawn between religion and technology in Accra.

Can Spiritual Entities Traverse Electronic Links?

Alexander's vision of the virtual utopia realized by combining the Internet and Christianity expressed how the Internet by itself was incomplete. The Internet alone could not answer the demand for social and economic transformation. An inability to realize expected gains through Internet use was not necessarily resolved through an adjustment in technical practices and was rarely attributed to a lack of technical proficiency. The reliance on religious practices to correct sociotechnical breakdowns indicated that users perceived the Internet as embedded within broader force fields of the material, social, and supernatural. The question of the Internet's relationship to spiritual forces was resolved in idiosyncratic ways by users who had diverse views about how to appropriately delegate their efforts between the material world and the supernatural realm.

Clarence, a student at a local technical school, spoke directly and critically about this matter of delegation, seeing Christianity as a force for progress that was often misinterpreted and overrelied on by churchgoers. He noted that "the church is a nice place to build the economy. Europe and America was built by the missionaries. . . . [the] church was used to build, they contributed so much to the civilization of this world." Clarence saw the historical role and the future potential of churches as forces for modernization. He expressed admiration for Mensah Otabil, founder of the International Central Gospel Church, one of the major megachurches in Ghana, whose pragmatic sermons address economic problems and underdevelopment focusing on what Clarence describes as "the real facts of life." Yet Clarence was critical of the way churchgoers in Accra placed all their energy in church to the neglect of work and school, observing that "Monday morning you're supposed to be at work or reading something, as early as 8 o'clock somebody is at the church praying that he wants to be rich. It's not magic like that. God doesn't work that way. That's a very big stumbling block. And if examinations are coming, you study, you don't pray. They drop their books, start going to churches, start going to conventions, crusades and they come and they fail their exams because they haven't learned. They want God to learn for them. . . . So I'm seeing Christianity as a weapon that is destroying us a little bit cause of the way people are approaching it for money. It's really bad." He argued that churchgoers too often relied on supernatural intervention and neglected to invest in the physical realm where Clarence, at least, was convinced hard work would be rewarded. Clearly these churchgoers had energy to expend toward their personal development but from Clarence's perspective they were delegating it poorly.

Clarence disputed the efficacy of addressing social and material breakdowns solely with adjustments in religious practice. He argued for the reverse image of Alexander's model of the incompleteness of the Internet. Clarence asserted that religious practice alone was similarly incomplete but (in keeping with the prosperity gospel) that it was connected to worldly prosperity—through its capacity to contribute to civilization. Alexander, perhaps a prime example of the behavior Clarence critiques, committed much of his time to prayer retreats and fasting rituals and he had gathered a group of individuals he called his "prayer warriors." He treated prayer as a tool for mediating between humans and the divine. He had established many contacts online who were the focus of prayer and his healing ministry. They had prayed for the father of a man he had met online who had cancer and were later told that the father had gone into remission. This notion of prayer as force is reflected in the encouragement he expressed in one sermon to "pray until something happens." This practice was seen as a way an individual might have an influence on worldly and otherworldly planes.

Alexander recognized a separate matter of delegation, not between the supernatural and the sociomaterial realm, but between virtual and physical spaces. Users like Alexander, who were exploring this intersection between religious and technological practice, held conflicting views about whether a distinction ought to be made between forms of human interaction: face to face, mediated by phone, or mediated by chat room. Some distinguished between the impact of synchronous versus asynchronous communication on the effectiveness of spiritual practice whereas others did not. Alexander recognized that the effectiveness of prayer had a limited reach, that one can pray only for known individuals with known problems. The Internet greatly expanded his ability to collect known individuals from around the world. Alexander also sometimes conducted prayers with chat partners via keyboard but did not assert that there was an effect of this synchronous prayer between distant individuals beyond providing social support. Yet there was value attached to simultaneity itself and he implicitly attributed importance to the timing and location of these prayers, commonly conducting them late at night and in more sacred places outside of the urban center. According to Alexander the simultaneity of ritual between the one praying and the one that is being prayed for facilitated shared access to a virtual space—not cyberspace but rather the "realm of the spirit." As he instructed, "I just say well I'll be praying for you today. Sometimes I do give them directions, pray tonight around eleven o'clock and *I'll be meeting you in the realms of the spirit.* So by eleven o'clock I will also be praying, so together we will be praying" [emphasis added]. Much like other Internet users who hoped to progress from online introductions to phone calls to in-person meetings, for Alexander, the Internet did not serve as an equally sufficient space for spiritual practice but rather as a launching point for more authentic, effective, and dedicated spiritual practice. His offline activities—giving church sermons, fasting, and prayer rituals—remained a necessary and vital part of his faith and a mechanism for effectively directing spiritual energies.

The practice of addressing pragmatic material and social problems with appeals for supernatural intervention is indicative of a particular metaphysics in which the sociomaterial and spiritual are closely intertwined. Related to this perceived intimacy between the worldly and otherworldly, Piot argues for an African conception of the self as fundamentally relational. Sounding very much like an actor-network theorist, he describes "the person as composed of, or constituted by, relationships, rather than as situated in them. Persons here do not 'have' relations; they 'are' relations" (Piot 1999, 18). Piot adds, "not only is the self in these societies tied to other human beings; it is also diffusely spread into the nonhuman world of spirits and ancestors" (Piot 1999, 19). This closeness between the human and nonhuman was reflected in the way spiritual entities were accounted for by some Internet users in the systems they built to realize economic and social gains such as education, migration, or business opportunities. The Internet (or the social networks it facilitated) were subject to the forces of these spiritual entities, good and bad.

One example of the way spiritual entities were accounted for was Gabby's attempt to enroll the help of a mallam from Nima to improve his ability to gain money from chat partners. At the time of our first meeting in late January 2005 he had already spent several months in chat rooms posing as a woman trying to lure a foreign boyfriend (the Chinese businessman) into a relationship. He had made no financial gains with this strategy and was growing frustrated. Gabby noted that many of his friends who were doing Internet scams had been aided by mallams, a role described as a healer or teacher who practices a syncretic blend of Islam and animist practices, so he sought this aid himself. He treated the religious options available in Accra as a buffet of choices detached from any moral framework, with each option presenting strengths and weaknesses. He commented that he sought the use of fetish to gain money from chat partners because he was in a hurry and "God answers prayers slowly." He compared the Koran and Bible matter-of-factly, noting that "I don't know why they are able to use the Koran to retrieve monies . . . because you cannot use the Bible for that." He describes these sacred texts in much the same way as one would expect technologies to be depicted: as mechanisms for instrumental gain.

One young mallam I spoke with, accompanied by Gabby, explained that he had adapted rituals taught to him by his grandfather to meet the demand for solutions to contemporary concerns such as obtaining travel visas, getting goods shipped from e-commerce sites (ordered with stolen credit cards), and gaining money from Internet chat partners. This mallam told Gabby that he was suffering from a spiritual block that was preventing him from making money off of his chat partners. Gabby identified his girlfriend (whom he subsequently broke up with) and an aunt as the figures responsible for this spiritual block. Drawing on witchcraft narratives Gabby and the mallam blamed closely connected female individuals for monopolizing a finite supply of life force, thereby preventing his success. He noted, suspiciously, that the aunt had traveled abroad twice already despite having only a minimal education, suggesting a disproportionate reward for her position in life. Therefore Gabby located his problems with operating the Internet successfully in another part of the broader system of social and spiritual relations. What in a disenchanted world seemed to be a technical or social breakdown ultimately was identified by Gabby and the mallam as spiritual.

To address Gabby's ineffectual attempts to gain money from chat partners the mallam provided his services at a cost of 350,000 cedis (approximately $38). He was given a talisman and was instructed to bury it under a large stone. He was also supplied with a potion to drink and bathe with twice a day for seven days. The mallam instructed Gabby to give *salaka* (alms) of 60,000 cedis to the poor who line up in front of a local mosque. He was to drink the potion just before he talked to his chat partners. He was warned that the potion would take effect only once he had verbally relayed the request for money to the chat partner by phone. For several months after visiting this mallam and following his instructions, Gabby had no luck in gaining money from chat partners so he sought the assistance of another mallam also recommended by a friend. The second mallam gave him a potion that he was instructed to spray on his hands just before he began chatting so that the effect would operate through his hands and the keyboard.

The two solutions provided by two different mallams indicated a lack of consensus about how to configure an engagement with the supernatural realm when the Internet was involved. The prescription of the first mallam reflected doubts about the capacity for religious practice to be carried out effectively via the Internet, although the telephone had become transparent enough as a mediator that it would suffice. There was an implicit understanding in this instruction that the realness of voice conversations took precedence over typed conversations, despite the fact that both conversations were mediated by technology and both were synchronous. The phone had become transparent whereas the Internet remained opaque. The second mallam recognized no such precedence assuming that supernatural forces could be mediated effectively through the keyboard. In this alternate practice chat conversations had a similar enough status to phone conversations or face-to-face conversations for a supernatural force to operate between those conversing. Similar to these two mallams, religious followers who are already accustomed to considering matters of mediation between the worldly and otherworldly realms are also likely to engage in a more careful and sophisticated consideration of the mediation between the real and virtual provided by new technologies (Miller and Slater 2000).

Not only in Accra but also in the broader West Africa region there is evidence that these hybridized material-supernatural systems are expanding their sphere of influence. This expansion was reflected in the way the mallam adapted rituals to the contemporary demand for travel visas and for obtaining products from e-commerce sites. These spiritual systems were perceived as extending far beyond the kinship networks that previously circumscribed their effective force. Alexander, who prayed for his chat partners, envisioned his prayers as having an unlimited operating terrain. He and his prayer warriors could aid anyone in any place around the world. Supporting this extension as a regional trend, Smith has noted a change in ritual practice pointed out to him by a Nigerian Presbyterian minister. The minister noted that ritual killings (viewed as indigenous and satanic) were beginning to be used effectively against strangers whereas before they could be committed only against kin (Smith 2001). This expansion of terrain has accompanied a growth in connections with the wider world through domestic and international migration trends, broadcast media, and technological advances such as the Internet. This was not only true of Islam and Christianity, which were already internationally networked religions widely practiced in Ghana, but was also true of indigenous animist beliefs.

The symbolic significance of places such as London and America was such that local gods could no longer be so local if they were to address the aspirations and imaginations of their potential clients. In an article on antiwitchcraft cults, anthropologist Jane Parish described fetish priests who advertise the gods who possess them in terms of their contemporary capabilities. She commented:

> Other priests bragged that their gods knew all about world affairs, and they often competed with each other over how many countries their gods had visited. One priest said, "I am possessed by a god who speaks many languages including French and German. He has been to these countries and flew like a jet, . . . an army plane. . . . He flies around the world and has been to London many times. . . . Ask me questions about London. . . . He will know the directions to your home. . . ." According to another, "The god visits many different places. . . . He is very rarely at home. . . . He may be in Holland or America. I ring a bell here and he will come to the shrine immediately. He spies things everywhere . . . he sees a lot in different countries. . . . In London he tells me it rains with cats and dogs" (laughter). (Parish 2003, 26)

Like these fetish priests, the mallam in Nima described his influential role in a supernatural system whose efficacy was not circumscribed by kinship networks, location, or nationality. His brother who was living in Spain had requested help in getting a Spanish woman to fall in love with him. To do so the mallam created a diagram in sugar ink on paper divided into nine sections and containing Arabic references from the Koran. He then mailed the diagram to his brother by postal mail, who on receiving it washed the ink off with water and drank this watery sugar ink as a love potion. The mallam noted that he had a number of clients living abroad in Spain and the United States, a situation made more feasible by new communication technologies. He noted that he was in regular contact with his brother via email.

The mallam made no distinction between the effect of these practices on foreigners versus locals. A Spanish woman could be won over with a love potion just as a Ghanaian woman could. Furthermore, the love potion would still have its effect even when the ritual was bifurcated considerably in time and space. The expansion of terrain on which supernatural forces operate constitutes a globalization of animist-syncretic traditions. It can be interpreted as a response to the increasing number of Ghanaians who live abroad along with the improvement of communication technologies linking those abroad with the homeland facilitating these transnational rituals. As Ghanaians living abroad continue to have problems with work and love, there is likely to be some demand for the services provided by mallams, prophets, and others in similar updates to their long-established roles (Parish 2011). This expansion of terrain calls into question interpretations of local religious rituals as immutable carriers of tradition in opposition to modernity and to globalization. In certain domains, particularly where religious services were embedded in the money economy and in transnational networks, religious innovation flourishes.

The embedding of religious ritual in the market in Ghana has therefore yielded forms of religious entrepreneurship. Rather than sitting by as new forms of Christianity lay claim to their clientele, the practitioners of indigenous animist beliefs have instead flexibly redirected and reframed their services to challenge religious interlopers. It is a further example of how "the practice of mystical arts in postcolonial Africa . . . is often a mode . . . of retooling culturally familiar technologies as new means for new ends" (Comaroff and Comaroff 1999, 284). The underground, informal circulation of fetish objects and animist rituals emerges as a new world religion surreptitiously piggybacking on diasporic movements rather than arriving front-door style through proselytizing missionaries and televangelists. These practices insinuate themselves in the gaps between reality and aspiration. The supernatural realm is conceived of as a force field impossible for anyone anywhere in the world to escape. The travels of these migrant groups thereby effect a re-enchantment of the Western world.

Conclusion

Inside and outside of Ghana, a growing body of research shows that religious organizations, religious leaders, and committed followers find that old as well as new media technologies such as the Internet offer new capabilities that help them to better follow religious edicts and to fulfill religious missions (Miller and Slater 2000; Hoover and Clark 2002; de Witte 2003; Gueye 2003; Meyer and Moors 2006). In Accra, religious rituals and the metaphysics underlying such practices drew the attention of their practitioners in particular directions and therefore cemented a way of relating to other people and to the material world. There was an ongoing concern with individuals who may be blocking one's prosperity via witchcraft and with seeking divine assistance to make connections with those who can provide opportunities. The locally understood reality reflected in such belief was that one's wealth was deeply intertwined with one's social network, intensifying the drive to make contacts and build social capital, activities that were prevalent in the Internet cafés of Accra.

The compatibility of religion and technology in Accra was tied to the way both domains helped users work toward some of the same desired outcomes. In a sense, the Internet was merely the newest religion, competing with churches, mosques, and shrines for followers and offering some of the same tangible rewards. These rewards included migration opportunities, healing, monetary wealth, business success, as well as the opportunity for fellowship—to make friends, find a spouse, and develop a network of social support. Gabby, who sought assistance in scamming chat partners through the services of a mallam, represented how far this mechanistic view of religion could be taken. His comparisons—that prayer was slower than the use of fetish objects, that the Koran could be used for getting money but the Bible could not, reflect a totally demysticized and profane calculation of forces. Although not the norm, Gabby openly admitted to pursuing practices that others felt compelled to conceal. Public discourse in Accra was such that there was great pressure among modern urban Ghanaians to be perceived publicly as pious Christians (or if not Christian then Muslim). Yet, private behavior indicated a continuing belief in the efficacy and importance as outlined by indigenous cosmologies that predate these world religions and more generally a widespread interest in employing religious practices for tangible solutions to worldly problems.

The mechanistic view of religious practice was controversial. Ghanaians like Clarence as well as many religious leaders viewed a disproportionate focus on money as a misappropriation of faith. Similarly, when religious leaders focused on money it was sometimes perceived as an irreligious desire for profit with the church serving as an entrepreneurial venture in a way that was incompatible with morality and true faith. The embedding of religious institutions in the money economy in Ghana was an important influence on the way these institutions operated. The title of pastor was simultaneously considered high status and potentially lucrative in Accra. The relationship between religious leaders and followers was sometimes one of vendor-client in which money was paid to religious leaders for services rendered, sometimes even in proportion to how effective these services were.[7] Therefore these religious leaders were always under pressure to deliver results, to meet the requests (often financial) of their followers with the threat that they may turn to other churches, shrines, and rituals if these needs were not met.

Most Internet users were not as actively engaged in religious practices as a preacher like Alexander or Gabby who was in consultation with a mallam. Yet these cases demonstrated a rich counterargument to particular notions of how digital technology produces certain inevitable social changes, particularly its secularizing or democratizing force. What was broadly true of the religious orientation of Internet users in Accra was their shared view that a multitude of connections exist between religion and technology and, more generally, between religion and other aspects of the worldly domain. The particulars of this configuration between the spiritual and technological realms varied between the highly religious and the only weakly pious, but within these various configurations was a conviction that the mechanical and physical world alone was incomplete and that the supernatural realm could always undermine or facilitate its operation. Technology alone could not effect genuine social and economic transformation if it was against God's will or was interfered with by demonic forces.

The discussion of technology and religion concludes four chapters addressing local practices of Internet adoption and adaptation in Ghana. This examination has demonstrated, importantly, that the Internet café cannot be sufficiently understood by studying it as a bounded space. Internet users built connections to relate new technological forms with established social formations such as the religious practices addressed in this chapter. The interaction between user and machine was not the only or even the primary point where the technology was worked out and rendered functional by users. The circulations that made their way from the café to other institutions such as the church and back again were extended spheres of user agency and material constraint. Traveling along with these depictions, we are now far removed from a blandly instrumentalist version of the Internet's role in a place like Ghana. Uses of the Internet were guided not simply by pragmatic efficacy or subsistence needs but by a complex of motives related to reducing unpredictability and threat, reconciling local social relations, redressing global inequalities, and engaging the imagination. Thus far, this account has not yet confronted the specter of development that seems to pervade life in economically depressed African nations such as Ghana. This is the work of chapter 6, where emerging institutional models of digital technology's relationship to socioeconomic development are examined. In particular that chapter explores how links between the Internet and development was constituted in the discursive stew of a UN-sponsored World Summit that hosted a regional conference in Accra in February 2005 that I attended. This analysis is pursued by relating it to the now-established baseline understanding of the value and relevance of the Internet among young urban Ghanaians who inhabit Accra's Internet cafés.

6 Linking the Internet to Development at a World Summit

In recent years, the international aid sector has seized on digital and network technologies reframing them as tools of contemporary development practice. This chapter makes a brief diversion from the main thread of this book to consider this process. In an effort to ground such consideration ethnographically, I consider one particular event—the Africa regional conference of the World Summit on the Information Society (WSIS)—sponsored by the United Nations, which was held from February 2 to February 4, 2005, at the Accra International Conference Centre. This event took place midway through my initial nine-month period of fieldwork in Accra. It unfolded contemporaneously and in close proximity to the Internet cafés documented in preceding chapters. Juxtaposing these two social worlds, that of Accra's Internet cafés and of the WSIS proceedings, shows clearly the divergent ways their members came to understand the Internet, its capabilities, and its particular relevance in an African context.

At the WSIS regional conference, processes of document production, speech performance, and the choreography of conference events centered on information as an important social good. Information and its unencumbered circulation was depicted as critical to human and societal advancement. The WSIS championed universal connectivity as an imperative for the progress of developing countries. By contrast, in Accra's Internet cafés, participants came to understand the Internet through informal, small media circulations, in settings as diverse as churches, street corners, and school yards. The technology was not employed in this context for the purposes of neutral and impersonal information circulation. It was first and foremost a space of person-to-person communication where one's progress was facilitated or hindered by human gatekeepers. The Internet was perceived by these nonelite users in Accra's Internet cafés as a domain of contested representations and, at times, of what seemed to be exclusions motivated by race, geography, and nationality. The absence of overlap between the terms and categories that frame ideas about technology in these two milieus potently illustrates the profound disconnect between these social worlds.

Given the vocal claims by WSIS organizers about the inclusiveness of the conference proceedings, one could expect some number of Accra's Internet café owners, operators, and users to find their way into the conference hall. Yet this was not the case. I found that among the Internet café–going youth there was little to no awareness of the event, no sense that its inclusiveness was inclusive of them. The organized youth clubs of Mamobi considered in chapter 2 and especially the active and Internet-savvy Avert Youth Foundation would seem ideal candidates to prove claims of the conference's inclusiveness and yet they were not present. Only a small cadre from the elite entrepreneurial and managerial crowd affiliated with BusyInternet were on hand at WSIS events. The absence of these local, nonelite populations from the conference in spite of the convenient proximity and their apparent investment in the issue of ICT engagement thus demands further explanation.

For long-time observers of development institutional processes such a disconnect likely comes as no surprise.[1] The extensive international travel undertaken by officials of the large international aid organizations (particularly the UN and the international financial institutions [IFIs]) and their apparent effort to consult with experts, governments, and local groups belies an essentially inward orientation and self-referentiality in the way such organizations tend to operate. This is facilitated, in part, by practices of physical cloistering: to the air-conditioned conference centers, four-star hotels, and chauffeured private transport, practices that were very much in evidence at the WSIS regional conference in Accra. This cloistering clearly preserves the prestige of such work and protects its uninterrupted efficiency and claims of universality. In light of such practices, the decision to locate the WSIS regional conference in Accra had far more symbolic resonance than practical consequences. The event unfolded in a kind of placeless space marked by a standardized environment that bolstered the standardized procedures.[2] There was also the cloistering of authority to the confines of certain disciplines, discourses, and professional roles.[3] This has consequences for the way these bodies forward the possibilities of information and of information technologies through their routine practices of expert consultation.[4] Opportunities to incorporate ground-level realities are thereby thwarted even when they are so close at hand. For example, a ten-minute taxi ride would take a WSIS participant from the conference center to any number of local Internet cafés where a perspective on these technologies wholly at odds with the way they were characterized in conference discussions would have been quite apparent.

The observation of a disconnect is the starting point rather than the main argument for the analysis that follows. To argue that practices of the international aid sector are unable to accommodate local knowledge and local particularities is not a particularly new insight (Escobar 1992; Jasanoff 2002; Hill 1986; Easterly 2006). However, such institutions are also continually concerned with self-perpetuation and are far from permanently fixed and inflexible in their organizational procedures. The more routine criticism and well-established arguments made against these agencies must be revisited in the wake of the inevitable reform efforts of the organization under pressure from high-profile ex-members, powerful critics, and populist protests. Criticism of the paternalism and antidemocratic process of these institutions, in particular, has altered the way the institutions themselves establish and defend their logic of development intervention. Some scholarship has begun to consider this institutional repositioning (Elyachar 2002; Raboy 2004; Riles 2001; Mohan 2000). Likewise, the critique of development must keep pace with emerging, new ways of employing language and concepts in processes of negotiation, debate, and consensus building.

This chapter considers certain procedural innovations enacted at the WSIS Africa regional conference in Accra. In particular the language of "inclusion," references to the distinctive "multistakeholder approach," and the incorporation of "civil society" groups into the process were all widely publicized components of the new and improved World Summit process.[5] Additionally, this chapter considers the imperative to "reach consensus" at the WSIS and how this shaped the forms of agency granted to groups participating in the event. To do this I examine language use, especially the repetition and alignment among terms within and between WSIS documents. The production of documents was a primary goal and final outcome of the conference proceedings and thus the documents themselves (bundled together into the *WSIS Outcome Documents* [WSIS 2005]) warrant careful attention. I attend also to elements of the aesthetics of information—the patterns, frequencies, and designs of words and documents taking a cue from Riles's study of nongovernmental organization (NGO) involvement in the World Summit on Women in Beijing, China (Riles 2001). The materiality of the mode of transmission, such as the quality of paper and printing, of intermediate and informal documents, and final and official documents is also significant. However, this social world is not captured in its entirety in the documents alone. In addition, the temporal unfolding of the event and the arrangement of the conference center space and its surroundings were components of the contradictory performance of these reforms at the WSIS. This process is attended to as well in the following analysis.

Arriving at the WSIS Regional Conference

The WSIS regional conference in Accra drew a diverse group of politicians, aid workers, NGO representatives, journalists, students, academics, entrepreneurs, and business executives. No groups or individuals were explicitly barred from the public proceedings and gaining access I found was a simple matter of filling out and faxing a form. However, awareness of the conference, as I have noted, circulated primarily within certain elite social networks. Additionally, the setting communicated itself visually as a space for elites. The Accra International Conference Centre is a large, showplace facility complete with air-conditioning and tinted windows. One can see out of the building, but one can't see in. The center sits regally atop a hill on a sprawling piece of land lined by a tall fence, a view of the building obscured by large trees. Behind the large gates the driveway circles up to the entrance shaded by a carport. The property is landscaped with a green lawn, very unusual for dusty Accra, and dotted with carefully pruned, topiary-like foliage. My arrival by local taxi was somehow controversial, the guards at the gate did not want to allow the taxi driver inside and signs barred pick-ups and drop-offs on the street in front meaning that I had to make a bit of a trek to pick up a taxi on my way out. The circular driveway and apparent lack of a parking lot indicates the expectation that visitors will arrive by chauffeured private vehicle.

During the three days of the WSIS conference, participants circulated and mingled outside the entrance to the main auditorium, which was set up as a space for NGOs, businesses, and UN agencies to display information about their projects and advertise their services. In one hallway a makeshift computer lab had been set up where participants could check email. In the main auditorium each African UN member-state was allocated seating front and center, whereas on the right side local chiefs in traditional dress were seated as observers. Behind these seats were sections allocated to civil society, to the press, and to other groups (see figure 6.1). In the main auditorium high-status speakers including President Paul Kagame of Rwanda as well as President J. A. Kuffour of Ghana made appeals for an inclusive information society. Smaller conference rooms held workshops where special topics such as ICT and socioeconomic development and access and infrastructure were debated.

Through the World Summit process certain key, new themes are introduced and validated ideally to chart new directions in international

Figure 6.1
The auditorium at the Accra International Conference Centre set up for WSIS

development work. Such themes are those deemed most worthy of the intense investment required to carry out an event on such a global scale. They must be compelling enough to draw the involvement of the huge numbers of people required to make such events noteworthy and influential. However, the World Summit process is also carried out, as Riles notes, "in an atmosphere in which international institutions struggle to defend their legitimacy on many fronts" (Riles 2001, 7) and thus may play an important role in publicly demonstrating progress and reform efforts within the United Nations in response to critics. To defend this legitimacy organizing institutions have offered a (seemingly) expanded role in conference proceedings to local groups and to grassroots organizations within developing regions. This is an effort to manage what critics call the *democratic deficit* in such institutions of global governance whose officials and representatives are not held accountable through direct elections (Willetts 2006, 308; Scholte 2002, 289). Although the intention to be more broadly inclusive and accountable are clear, how these efforts to expand participation actually play out in practice have not been well documented.

The official line of the World Summit series of conferences is that they have an impact by "serving as a forum where new proposals can be debated and consensus sought" (United Nations Department of Public Information 1999). The question is how consensus building and debate is acted out by conference participants in this context. One could find an element of conflict (and therefore possibly debate) in relation to the caucuses present at WSIS championing the causes of youth, women, and civil society. These caucuses presented themselves as minority groups fighting to have their voices heard. A particular example illustrates the contradictions in this performance. On the second day, just outside the main auditorium, I noticed a folding table on which loose-leaf papers had been scattered across. These papers were copies of a one-page document produced by an organization identifying itself as the African Civil Society for the Information Society (ACSIS). The short statement printed on these papers voiced their objection to what they perceived as exclusion from conference proceedings. In stating their case, they employed UN linguistic conventions noting:

> Considering the goal of the international community in building an inclusive information society in a multi-stakeholder framework; Government, Private Sector, International Organizations, and the Civil Society.

We, the *African Civil Society*, deeply regret our exclusion from the opening ceremony. Not only did we find it difficult to gain access to the main hall, we were denied a speech slot, thereby barring us from bringing the grassroots' message to the conference.

Some details about the format of the document itself are pertinent. To demonstrate its effectiveness as an informal document of a grassroots organization it can be placed alongside a similar document—an agenda I picked up at a meeting of the Avert Youth Foundation in Mamobi (see figure 6.2). Avert Youth Foundation was, by any definition, a civil society group—perhaps precisely the kind of group WSIS organizers wished to conjure up for the media and the public when they talked about inclusion. Avert was an organized body composed entirely of youth living in the Mamobi neighborhood. It had elected leaders, a constitution, regular meetings, and organized activities oriented toward the common good of youth and the broader community of the neighborhood. There are certain

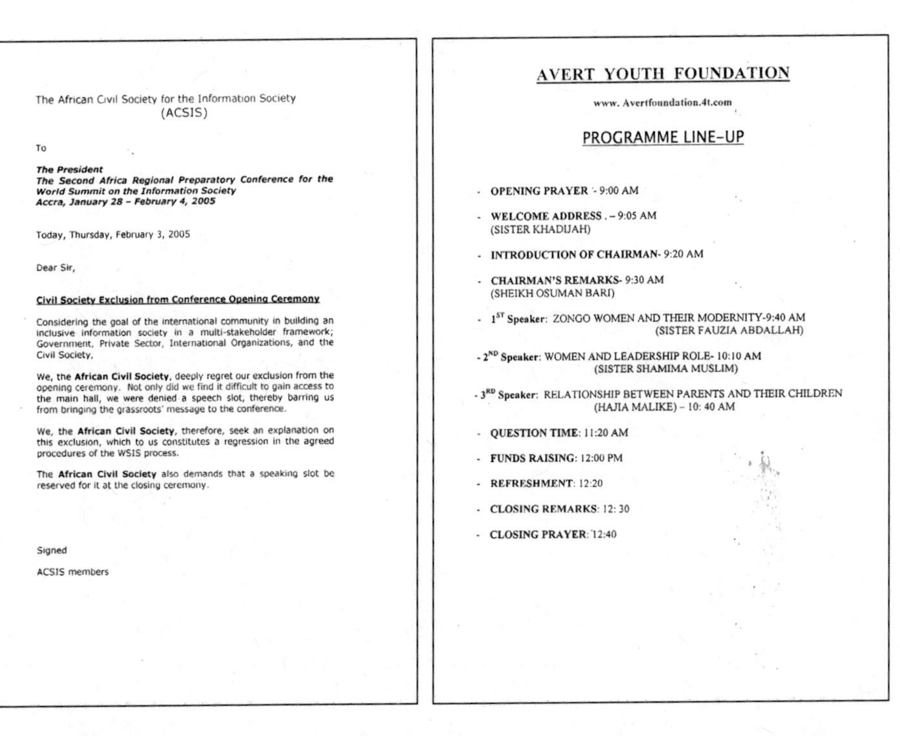

The African Civil Society for the Information Society
(ACSIS)

To

The President
The Second Africa Regional Preparatory Conference for the World Summit on the Information Society
Accra, January 28 – February 4, 2005

Today, Thursday, February 3, 2005

Dear Sir,

Civil Society Exclusion from Conference Opening Ceremony

Considering the goal of the international community in building an inclusive information society in a multi-stakeholder framework; Government, Private Sector, International Organizations, and the Civil Society,

We, the **African Civil Society**, deeply regret our exclusion from the opening ceremony. Not only did we find it difficult to gain access to the main hall, we were denied a speech slot, thereby barring us from bringing the grassroots' message to the conference.

We, the **African Civil Society**, therefore, seek an explanation on this exclusion, which to us constitutes a regression in the agreed procedures of the WSIS process.

The **African Civil Society** also demands that a speaking slot be reserved for it at the closing ceremony.

Signed

ACSIS members

AVERT YOUTH FOUNDATION

www. Avertfoundation.4t.com

PROGRAMME LINE-UP

- **OPENING PRAYER** - 9:00 AM
- **WELCOME ADDRESS** . – 9:05 AM (SISTER KHADIJAH)
- **INTRODUCTION OF CHAIRMAN**- 9:20 AM
- **CHAIRMAN'S REMARKS**- 9:30 AM (SHEIKH OSUMAN BARI)
- 1ST **Speaker**: ZONGO WOMEN AND THEIR MODERNITY-9:40 AM (SISTER FAUZIA ABDALLAH)
- 2ND **Speaker**: WOMEN AND LEADERSHIP ROLE- 10:10 AM (SISTER SHAMIMA MUSLIM)
- 3RD **Speaker**: RELATIONSHIP BETWEEN PARENTS AND THEIR CHILDREN (HAJIA MALIKE) – 10: 40 AM
- **QUESTION TIME**: 11:20 AM
- **FUNDS RAISING**: 12:00 PM
- **REFRESHMENT**: 12:20
- **CLOSING REMARKS**: 12: 30
- **CLOSING PRAYER**: 12:40

Figure 6.2
Informal documents of two civil society groups

parallels between the documents of ACSIS and the Avert Youth Foundation. Both are printed in plain black ink on a single sheet of paper. Both show small blotches of toner indicating low-quality photocopying. The format of the medium supports the claims of ACSIS to grassroots and civil society identification. The discovery of the single-page appeal unbound from any document communicates also the group's marginalization.

Yet in the document from ACSIS it was clear not only from the UN resolution–style format of this statement, but also the use of certain terminology—*information society* and *multi-stakeholder framework* in particular—that this debate was taking place on pre-established discursive grounds. In this sense the grassroots message of this caucus did not challenge the orientation of the UN and WSIS organizers toward technology on any fundamental level but ultimately reinforced its claims. The various caucuses present at the WSIS regional conference produced numerous informal documents of this sort. These documents uniformly confirmed that indeed the information society was a salient reality for these caucuses. Its importance was again and again reasserted through their demands for inclusion. As a newsletter published by a caucus of allied youth stated, "sessions after sessions have been passionate about the powerful effect of ICT in addressing our socio-economic challenges, circumstances and aspirations. ICT is inspiring; isn't it? Indeed ICT has come to stay and for our generation or the youth to effectively use and fully participate in the information society, it is imperative to *talk back* over and over again on the inclusion solution—Access" [emphasis added]. The youth newsletter and African Civil Society memo employed language that portrayed their group's role as adversarial to conference organizers and participating governments and yet the message from these groups was far from controversial. Despite official claims for diversity and debate, these documents demonstrate how participants at WSIS spoke in one voice on certain core subjects regarding the Internet, development, information, and the information society.

Why Hold a World Summit on the Information Society?

In this section, stepping back for a moment from this immersion in conference minutiae and the jockeying for position between conference participants, I seek to explain how the concept of an *information society* came to be selected to mobilize so many individuals, groups, and nations. Why did this particular concept rise to such prominence at this particular point in history? To get at the utility of the term in the context of global governance and international aid it is helpful to review a bit of the recent history of the development industry. The WSIS Africa regional conference in Accra gathered together representatives from participating African states as well as a number of attendees from the private sector, from various nonprofit groups, and from the organizing bodies, the United Nations in partnership with the International Telecommunication Union. Unlike previous World Summits in which a single conference was the culmination and focal point of preparatory activities, the WSIS was organized into two parts: an initial conference in Geneva in 2003 followed two years later by a final conference in Tunis, Tunisia, in November 2005. The WSIS conference placed the information society alongside other World Summit themes including human rights, gender equality, and environmental sustainability as topics with substantial cachet and widespread public interest inside and outside of development circles.

The way the concept of an *information society* is positioned in the *WSIS Outcome Documents* (WSIS 2005) illustrates certain alliances that those participants involved in drafting the documents sought to realize with the term. The *WSIS Outcome Documents* include the "Geneva Declaration of Principles" and the "Geneva Plan for Action" from the 2003 conference and the "Tunis Commitment" and "Tunis Agenda for the Information Society" from the 2005 conference. The phrase *information society* appears most prominently in the outcome documents as a way of framing calls for inclusion, as in the assertion at the beginning of the Geneva "Declaration of Principles" that "everyone, everywhere should have the opportunity to participate and no one should be excluded from the benefits the Information Society offers." The information society is thus depicted as a kind of international club, a collaborative body, a community in the most idealistic sense where "everyone can create, access, utilize and share information and knowledge" toward "sustainable development" and "improving their quality of life" (WSIS 2005, 9). There is no indication of any inherent trade-offs or downsides to this new global order except in the acknowledgment that certain groups are currently excluded from membership.

The concept of the information society originates in the academic literature that also includes a number of proximate terms: the *postindustrial society* (Bell 1974), the *information economy* (Porat 1977), and the *network society* (Castells 1996). The concept definition work these authors undertake in each case centers on a postindustrial shift in the principal wealth-generating center of some national economies and in the emerging global economy. The shift is from the manufacturing of commodities to service work, data processing, financial services, and knowledge production and management (Webster 2002). The WSIS documents attempt an extension and realignment of the concept with the themes of development, equity, and poverty eradication, all concerns mostly absent from the source literature. Thus in the process of borrowing the information society as a World Summit theme, the concept was substantially altered and redefined according to the needs and interests of involved organizations. The WSIS documents, as one would expect, are normative in tone and encompass a wide variety of socioeconomic goals, equity issues, and stakeholder groups. They downplay the transitions in national economies as well as in the global economy that scholars typically associate with the information society. Instead the central focus is on unencumbered information circulation as the key to realizing progress for all of humanity, including poor, marginalized, and vulnerable populations.

In the *WSIS Outcome Documents,* the significance of specific technological artifacts or systems is largely subsumed by the concept of the information society. The category "Information and Communication Technologies" (abbreviated ICTs) is used frequently throughout the documents. ICT is an abstraction meant to encompass some set of distinct technologies, though they are never actually identified and enumerated. In the WSIS documents ICT is treated as the material embodiment of new opportunities and as a kind of tangible portal into the information society. The category ICT is sometimes employed as a real and singular entity as in the call for "ICT public access points in places such as post offices, schools, libraries" (WSIS 2005, 15). Among the possible instances of ICT, the Internet is the one that appears most frequently, though open source software is another more specific reference. In the WSIS documents the Internet appears alongside calls to ensure access. It is depicted also as an entity that must itself be governed, ideally collaboratively and equitably, among regions and governments. A ten-page section in the "Tunis Agenda for the Information Society" follows the subheading "Internet Governance" (WSIS 2005, 75–85). There is also one brief mention of the continuing relevance of radio and television (WSIS 2005, 88). The way specific technologies are (and are not) referred to in the WSIS documents reflects a certain ambiguity around their particular capabilities and possible applications. The use of the term *ICT* disengages from the diverse and divergent materiality of the various technological forms that the term seems intended to encompass—including the Internet, computer hardware and software, as well as mobile phones and older technologies.

The particular alignments organizers and participants sought to realize through the WSIS process are visible in these outcome documents. A sense of why the "information society" joined the other key development concepts in the World Summit series precisely when it did can be deduced from the broader context of UN reforms and trends in the business community, including the dot-com boom and its aftermath. The WSIS was initiated by a UN General Assembly Resolution in 2002, just as the dot-com boom had begun its downward spiral. Yet the seemingly transformative effect of this newly emerging industry and its revitalization of the high-tech sector, the astounding wealth generation that accompanied it, and the highly publicized availability of capital embodied by the figures of the venture capitalist were likely on the minds of many governments as well as international bodies such as the United Nations at this time. What also preceded the initiation of WSIS was a notable shift in the private sector toward a broader social role for the corporation (Reich 1998). Evidence of this is found in the increasing prominence of corporate social responsibility efforts, social investing, and corporate foundation work. A more purely profit-oriented interest in these firms considered the possibilities of the new consumer markets represented by emerging economies particularly in the countries grouped as BRIC (Brazil, Russia, India, and China). A direct appeal by then secretary-general of the UN, Kofi Annan, to "Silicon Valley decision makers" was published November 5, 2002, on the *CNET news* Web site (Annan 2002). In this article he calls for the broadening of corporate social investment beyond the United States to international issues and specifically requests the participation of members of this regional industry in the upcoming WSIS events.

At the same time, as the 1990s drew to a close, the domain of international aid work was at the tail end of enthusiasm for neoliberal reforms and structural adjustment programs with a consequent nadir of state power in many developing countries.[6] The vacuum this created, particularly in providing certain critical social services, was increasingly filled by other organizational forms, especially by NGOs. Consistent with neoliberal thinking, these organizations were conceived of as a way to bypass the institutional weaknesses of corrupt governments or failed states (Smillie 2006; Elyachar 2002; Fisher 1997). The initiation of the WSIS process was concomitant with a growing awareness within the UN of the many other players emerging in this space that could perhaps be enrolled in ways that contributed to the UN's institutional legitimacy. Two years before the WSIS process was started, Annan launched his "Global Compact" at the World Economic Forum in Davos in January 1999 with the remark, "The United Nations once dealt only with Governments. By now we know that peace and prosperity cannot be achieved without partnerships involving Governments, international organizations, the business community and civil society. In today's world, we depend on each other."[7] The Global Compact sought to align businesses with UN principles in the areas of human rights, the environment, anticorruption, and labor standards. In committing to these principles companies became an enactor and enforcer of goals set by UN agreement, though on a wholly voluntary basis.

Organizational shifts within the UN are often marked by the appearance of new terminology and such is the case here as well. The term *public-private partnership (PPP)* emerged to describe, in positive terms generally, the possibilities of an alliance between the private sector and one or more government agencies, an apparently postneoliberal reconfiguration that highlights collaborative dimensions of this relationship. Thus such a term appeared, at least rhetorically, as distinct from the more absolute transfer from public to private marked by the term *privatization*. Public-private partnership has since become a prominent concept in describing efforts to realize broader access to ICTs in developing regions (Kuriyan and Ray 2009; Fife and Hosman 2007). This new prominence of for-profit businesses in development promoted by UN officials is also reflected in the *WSIS Outcome Documents* in which references to the "private sector" are frequent and often appear in lists alongside governments, civil society, and international organizations as parallel stakeholder groups. The term *public/private partnership* appears several times as well in the WSIS documents. By redefining the information society as a global community and making this the unifying concept of a World Summit, organizers and participants pursued an alliance among diverse new groups including the private sector, particularly the high-tech industry with its deep pockets, but also as a counterbalancing measure, with NGOs that typically claim a certain populism and community-alignment in their interests and activities.

This opening for private sector involvement and the boom in the hardware, software, and service industries associated with the Internet partly explains why a World Summit on the information society took place when it did. However, it is also worth considering why the initial personal computer revolution did not generate a similar alliance and why the information society as a key concept in international development emerged instead in the second wave of innovation sparked by the Internet. How is it that a certain social idealism tied to advancements in human well-being beyond new jobs, industries, and an improved economy came to be attached to digital network technologies? One explanation might be a materialist one, that the earlier personal computing revolution made possible a revolution in business and labor productivity whereas the capabilities of networking and the Internet were required to finally accomplish more fundamental and universal possibilities for social connection and ultimately human liberation. By contrast, a credible nonmaterialist history is offered by Fred Turner's study of the countercultural origins of the Internet and cyberculture (Turner 2006). He details how technology enthusiasts began to forward this perceived compatibility between computing technologies and human self-realization as far back as the 1980s wave of innovation around the personal computer. At that time such a connection was only nascent, driven in a large part by hobbyist communities. By the second wave of innovations around the Internet, the connection between the new network technologies and ideals of self-realization, collaboration, and communalism finally came to full fruition attaining an entrenched place in the public imagination. This, Turner observes, was in no small part due to the way the envisioned liberation of human capacities via information and networked connection was publicized in mass media outlets and in the especially sympathetic new venue offered by *WIRED* magazine. Such ideals were consumed by the public and the emerging industry itself. They were further enhanced through an emerging business culture in which entrepreneurs and executives propounded the egalitarianism of their organizations and the liberatory ideals underlying their products and services while rendering concrete these ideals in the demonstrated profitability of such ventures.

This new ethos of the Internet was, as it turned out, also quite compatible with the evolving orientation of the international aid sector. It was further concretized and materialized by intermediaries that moved between these worlds. Here I wish to consider the emergence of a number of entrepreneur-philanthropists as a new type of actor in the development industry emerging around the turn of the millennium. Some of these individuals had made a lifetime's fortune in the dot-com boom and were looking to formulate a meaningful second act to their lives. I became aware of two such individuals through organizations that had a presence in Accra. GeekCorps, an NGO modeled after the Peace Corps, was organized to bring technically skilled volunteers from the Global North to a small set of countries in the Global South. Through these volunteers, the organization sought to provide assistance and training on Web site design, network management, and related skills to small and medium-sized businesses. GeekCorps was started by Ethan Zuckerman who had founded Tripod, a Web-hosting service that he and his cofounders sold to Lycos in 1999. Later, after GeekCorps became a program of USAID, Zuckerman followed a second project organizing the nonprofit Global Voices, an officeless virtual community of more than two hundred Webloggers worldwide focused on bringing forward "voices that are not ordinarily heard in international mainstream media."[8]

A similar story is behind BusyInternet, the large one-hundred-screen Internet café, conference center, and business incubator in the center of Accra, Ghana, and one of the fieldsites for this research. A team of three business partners including Welsh-born entrepreneur Mark Davies founded BusyInternet. Davies had built his reputation and his fortune from the online city guide *Metrobeat* (that later merged with *CitySearch*) and a venture capital networking forum called First Tuesday. Zuckerman and Davies both had spent time in Ghana—Zuckerman as a Fulbright scholar, Davies as a tourist—prior to pursuing technology projects there. An inveterate traveler Davies sought to bridge this interest with his technical background. He commented vividly, "I had always been interested in Africa . . . and I obviously was interested in the Internet and I was sitting on the back of a bus in Copacabana in Rio wondering how to combine these two interests." He described his own introduction to the Internet as a personal epiphany noting, "I remembered the night that I sat in front of this thing and the first time that I discovered the Internet and I just sort of like, my mind just sort of like . . . exploded with excitement and possibilities and ideas. So I thought well, you know, I've made money in America and I know other people who've made money in America. Why can't I get them to all give like $50,000 or something and we'll go off and we'll set some sort of philanthropic access center with like thirty PCs and provide access." Such social investors saw the new network technologies not simply as a means to accomplish conventional development goals or as mere tools of profit generation but as embodying certain liberatory ideals and as a means of empowering societies, enabling self-expression and the greater visibility of marginalized groups in the global community.

Ventriloquism

In my fieldnotes and interview transcripts, I have no record of the term *the information society* ever being uttered by any of the Internet café owners, operators, and users I encountered in the course of fieldwork. Such a concept was wholly absent from the Internet café scene. Specialized language is often used to mark the boundaries of social worlds distinguishing insiders and outsiders. In this case, divergent conventions in language used to talk about technology reflects something about the nature of inclusion at WSIS, how and on what terms participation was extended to grassroots groups and a conflicting sense of the permeability of the development institutions boundaries. Returning to the unfolding proceedings of the WSIS conference in Accra, there is the matter of how organizers laid claim to and presented their multistakeholder approach and framed what a civil society group was and its important, new role. This definitional work in turn shaped the way civil society groups at WSIS presented themselves as pot-stirrers and as champions of the marginalized.

In the foreword to the *WSIS Outcome Documents,* Yoshio Utsumi, secretary-general of WSIS and of the ITU, notes, "The Summit has been notable in its adoption of a multi-stakeholder approach, and this is now carried forward in the implementation phase with the direct involvement of civil society and the private sector alongside governments and international organizations" (WSIS 2005). By emphasizing direct involvement he constructs an image of an increasingly diverse body of participants wandering the conference halls and making their voices heard, no longer limited to the mediation of representatives. The explicit emphasis on inclusiveness was present at the initiation of the event, explicitly noted in the UN Resolution that launched the Summit, which encouraged "NGOs, civil society and the private sector to contribute to, and actively participate in, the intergovernmental preparatory process of the Summit and the Summit itself" (United Nations 2002, 2). Notably, in the compiled *WSIS Outcome Documents* the term *NGO* was dropped entirely in favor of *civil society*. It is worth considering the implications of this. There is a general consensus that NGOs constitute a subset within the category of civil society (Smillie 2006; Willetts 2006; Carothers 1999–2000). The notable swapping of terms at WSIS, shifting up a level from one category to a broader one, effects inclusiveness through language alone. However, this does not necessarily equate to alterations in the actual composition of World Summit participants.

Within the *WSIS Outcome Documents,* the initial "Geneva Declaration of Principles" from 2003 and the "Tunis Agenda for the Information Society" of 2005 contain the identical assertion that "civil society has also played an important role on Internet matters, especially at the community level, and should continue to play such a role." This is suggestive without being altogether too definitive. The notable phrase in this statement is "at the community level," which avoids committing too narrowly to the idea of a civil society that is of the community and still suggests that this is where it principally operates. Carothers argues that skepticism toward the equating of civil society with local, grassroots activism is warranted in part because civil society is also inclusive of "Western groups projecting themselves into developing and transnational societies. They may sometimes work in partnership with groups from those countries, but the agendas and values they pursue are usually their own" (Carothers 1999–2000, 27). The extent to which the civil society category was made up of such groups at WSIS compromises the vision of direct involvement. The WSIS Web site claims that 6,241 participants from "NGO and civil society entities" were present at the final Tunis conference.[9] However, the distribution between different types of groups is uncertain and thus the matter of how direct or indirect the participation of civil society especially among those "at the community level" at WSIS is not resolved simply by this impressive show of participation.

In practice, the registration process for the WSIS conference in Accra facilitated an extremely loose and broad definition of what might count as a civil society participant. Thus the participation numbers for this category were perhaps higher than they might be otherwise.[10] The civil society became a kind of residual category composed of any individual who simply did not fit under the other more stringently defined categories. These included individuals who were not necessarily part of any organized group and did not have to be affiliated in any way with an organization located on the African continent. I myself was grouped under the label *civil society*, which was printed prominently on my WSIS nametag. Yet, presumably important stakeholder groups such as Accra's young Internet café users or the café operators and owners were not in attendance at all.

In addition to the actual composition of the civil society at the WSIS regional conference, there was also the matter of how the participants grouped under this designation made themselves heard. Certain constraints shaped their voices within the conference proceedings. As Riles notes, the legitimacy of the UN and its agencies is broadened in scope as the number and diversity of participants grows (Riles 2001). Outside groups also treat their inclusion in UN summits as validation of their existence and of their specific causes (Riles 2001). Ultimately these groups must balance the sense that their role is to challenge authority with the desire to maintain access and to continue to be included. A formal WSIS accreditation process put this access at stake for civil society groups. This process was not required for attendance but did create varying degrees of inclusion at the conference. Accreditation involved an evaluation of "the relevance of the work of the applicants on the basis of their background and involvement in information society issues" (WSIS 2003). This demonstrates how a pressure to conform to the institutional framing of the issues came to be formalized in procedure. For outside groups seeking to move inside the WSIS body, adopting the institutional language was the price of their admission and participation.

In the case of the African Civil Society for the Information Society and their memo of protest against WSIS conference organizers as introduced earlier in this chapter, the formal and institutional language they rallied around their claims for inclusion was quite at odds with their public performance of conflict against this very same institution. The oddness of this way of speaking was most visible in the comparison with more unencumbered civil society organizing—of the youth of the Avert Youth Foundation, the youth bases in Mamobi, and even the Internet scammers—and the ways they viewed and made sense of the Internet. What was demonstrated by ACSIS was a form of speaking whereby grassroots, local groups are invited to make a statement but seem to do so in the voice of the development expert. It can thus be described as ventriloquism. Ultimately this practice served the aim of reinforcing the significance of the information society as a development concept and of orthodox notions of development but in a voice that seemed to emanate from the potential beneficiaries themselves. Therefore this new yet more sophisticated rhetorical form that was part of the multistakeholder approach, although perhaps a rhetorical innovation, did not provoke any profound institutional change in the processes of setting development priorities, building consensus, and identifying solutions at WSIS because it offered no space for truly counterdiscursive debate.

That WSIS participants on the one hand and Internet users at a grassroots level on the other literally speak a different language about technology to a certain extent supports Castells's argument for the existence of multiple networks that are distinguished from one another via their "communication codes" (Castells 1996, 501). Castells notes how the logic of the network is such that a failure to conform to these codes results in an immediate exclusion from the network (Castells 1996). Therefore, it is a logical effort on the part of groups such as the African Civil Society for the Information Society to perform these codes in order to maintain membership, be relevant, and have an impact on the proceedings, though through the process the scope of impact becomes increasingly limited. Castells warns that the "presence or absence in the network and the dynamics of each network vis-à-vis others are critical sources of domination and change in our society" (Castells 1996, 500). Therefore the disconnect between WSIS participants and Internet users in Accra amounts not just to wastefulness and missed opportunities for collaboration, but also has the potential to be realized as a competition in which one network (likely the one composed by powerful development institutions and the state) undermines in various ways the other.

New procedural practices at WSIS events are a response to critics of the UN and of development processes in the high modernist tradition (Scott 1998) more generally that are characterized by a top-down, technocratic approach. The multistakeholder approach is one response to this particular type of condemnation and concomitant calls for a more participatory process. However, in this case, the approach offered an incomplete and superficial response. It attempted to dispel criticism while still maintaining the status quo. In fact, the participatory approach that emerged during WSIS in a certain sense makes it even harder to criticize institutional elitism because it involved these performances of direct involvement by groups labeled (and labeling themselves) as the civil society. Critics who argue against such institutional processes as WSIS on behalf of marginalized groups come to appear in opposition to the groups they mean to champion. This superficial appearance of inclusiveness, where who exactly is involved in the World Summit proceedings remains obscured, may set back rather than progress efforts to push for more truly bottom-up processes for development innovation.

Alliance Building

The multistakeholder approach was meant to bring civil society as well as the private sector into the WSIS process as allies in the effort to promote the information society as a key development theme. During the WSIS process, as with all other World Summits, the practice of referencing allies served as a way of underlining the facticity and importance of claims made about technology and its link to development. There are certain parallels between this practice and the way Internet users employed allies in telling rumors about Internet fraud and scams (see chapter 4). Such rumors were often framed by the comment, "a friend told me . . ." or "I heard it on the radio." At WSIS, allies were often created out of authority figures (a traditional ethos form of rhetoric) or, in lieu of the powerful, by pointing to the massiveness of these allies with precise numbers. Consensus was more abstractly evoked using phrases such as *universally acknowledged* or *widely recognized.*

Several documents produced in conjunction with the WSIS process benefited from a preface written by former UN secretary-general Kofi Annan himself (Stauffacher et al. 2005; Stauffacher and Kleinwächter 2005). This was done as a lead-in placing the powerful alliance at the front of the document where it is difficult for readers to miss. The WSIS regional conference used the same approach putting powerful leaders (presidents, ambassadors, etc.) on stage in the main auditorium, the most visible location possible. By combining these strategies, using references to authority and to large numbers of allies, the rhetorical effect was multiplied, as in the statement from the WSIS Web site that "nearly 50 Heads of state/government and Vice-Presidents and 197 Ministers, Vice Ministers and Deputy Ministers from 174 countries as well as high-level representatives from international organizations, private sector, and civil society attended the Tunis Phase of WSIS and gave political support to the Tunis Commitment and Tunis Agenda for the Information Society that were adopted on 18 November 2005. *More than 19,000 participants from 174 countries* attended the Summit and related events" [emphasis in original] (*Basic Information about WSIS—Overview* 2006). In the same way that the persuasive power of rumors relies on a reference to sources (e.g., "a friend told me"), these alliances produce a powerful argument for the factuality of the claims. Established in this way, they were that much more difficult to counter (Latour 1987). To mount such a challenge meant countering each of these alliances and in the case of WSIS this included fifty heads of state and vice-presidents, 197 ministers, and nineteen thousand total participants.

Riles also notes the concern with numbers as a register of impact in her observation of how documents were produced at (and in relation to) the World Summit on Women in Beijing (Riles 2001). This was a concern not only of conference organizers but of participants as well. She noted how the meaning of the words was often secondary to their patterning and frequency. Participants would count the number of times a key word appeared in the resulting document as a measure of their impact, a kind of informal content analysis. Of special importance was the repetition of words or phrases, not only within the document, but also borrowed from other documents to tie the documents together. This was another way of defining allies. Along these lines, a recurring practice in the *WSIS Outcome Documents* was to link ICTs and the information society to the Millennium Development Goals (MDGs) such as the statement that "the role of ICTs, not only as a medium of communication, but also as a development enabler, and as a tool for the achievement of the internationally agreed development goals and objectives, including the Millennium Development Goals" (WSIS 2005, 68). There are twenty-seven references in total to the Millennium Declaration and the MDGs in the *WSIS Outcome Documents*. These documents were heavily intrareferential, justifying priorities through an alignment with other institutional documents but not referencing entities outside of the institution. Besides the MDGs, there are also references frequently made to the Universal Declaration of Human Rights, the Charter of the United Nations, the Vienna Declaration, and the Monterrey Consensus. By making these references, the WSIS documents borrowed the legitimacy of these already negotiated texts and also lent a sense of the continuity and strength of UN agreements.

Alliances established among individuals and between documents had the effect of reinforcing the broader UN and its agencies and efforts asserting the legitimacy and relevance of various entities (from the Charter of the UN to Kofi Annan) and borrowing legitimacy already accumulated by these entities. Yet it was precisely through such efforts to maintain the institution that the disconnect was produced with what was external to the conference and institution, including the populations thought of as the eventual beneficiaries of such a conference. By building up internal connections, external ones were neglected. The references to other UN documents and the consistent repetition of terms were part of the communication code that participants could employ to perform membership. Ultimately, this meant that participation required a conversion process to make participants suitable for linking as they moved across the institutional border. This conversion was enacted for the sake of the coherence of the event and to accomplish some semblance of consensus.

Despite claims that the WSIS process was a space where "new proposals" could be "debated" the diversity of participants and the drive to reach consensus necessarily muted efforts to bring anything fundamentally new to light. Two unenviable and arguably incompatible tasks were set for WSIS as for all recent UN summits—to be as broadly inclusive as possible and at the same time reach a unified consensus at the end. As Riles notes, these tasks accomplish two important aims: broadening the legitimacy of UN institutions as well as creating a high impact discourse that although not legally binding would achieve an unquestioned veracity powerful enough to shape future policy and practice worldwide (Riles 2001). In this context it is quite an accomplishment that the *WSIS Outcome Documents* run to only ninety-seven pages. The reality of an event involving nineteen thousand participants from 174 countries is that much of what goes on within the event is intra-institutional maintenance and negotiation, linking and alliance building. This leaves little room for convoluting matters by, for example, permitting the conceptual premise of the event to be called into question. Civil society groups in their self-defined role of holding governments and institutional power accountable focused on more manageable goals, such as getting the phrase *human rights* (a term already well known to the institution and the subject of its own World Summit) incorporated into outcome documents (Bloem 2005, 101). Given these conditions, the structure of WSIS, as with previous World Summits, prevented the airing of any radically counterdiscursive grassroots messages about technology and development. A truly bottom-up process, without an overarching framework and imposed discourse for organizing the concerns of participants, simply would not function at an event of this size and scope. Thus, in the drive toward consensus the institution was forced to disconnect.

Conclusion

Conference organizers at the WSIS constructed and publicized an image of grassroots, community-based groups being ferried through the doors to comment directly on the process of priority setting and decision making. Yet the particular language used by the so-called civil society groups present at the WSIS reaffirmed an institutional discourse rather than challenging it on any fundamental level. These groups uniformly validated the information society as a unifying concept, even as they acted out opposition and conflict with conference organizers. The acting out of this conflict supported superficially the conferences' claims that they were open to debate and inclusive of diverse stakeholders. The institutionally affirming language of civil society groups at the WSIS reflected constraints on the agency actually granted to such groups, their conditional inclusion. Despite claims of inclusiveness, certain segments of civil society that were deeply engaged with the possibilities of digital and network technologies (e.g., those involved in Accra's Internet café scene) were not present at the conference at all. The WSIS consequently had little bearing or relevance on ICT engagement among nonelite youth in Accra who carried on in the wholly separate sphere of the Internet cafés.

The Internet cafés and their users remained invisible to the regime of development. The government of Ghana was, in fact, in the process of developing a competing network of community information centers (CICs) offering Internet cafélike access facilities in urban and rural areas of the country under the direction of the Ministry of Communications. A glossy pamphlet about the project was handed out at the WSIS conference in Accra. The political economy of the Internet in Ghana, as such examples illustrate, was not formed just by the governmental constraints and pressures that directly improved or limited access to technology, but also indirectly by the absence of intrusion into the Internet café scene. This was not automatically a negative turn of events. As long as the Internet cafés remained outside this regime they were available for the unencumbered claim-staking of youth as they devised their own scripts for this complex, multipurpose technology. Of course this freedom to reinvent came at the cost of whatever additional resources, rebalancing of access (beyond male youth), or broadening of possible uses that development institutions and the national government might have offered were they to treat the Internet cafés as part of their domain.

In analyzing the procedures and practices of a UN conference from the ground-level perspective of a participant, I have employed largely the same approach already applied to the study of Internet cafés and urban youth in Accra. The intent was to draw attention to the quite similar practices and performances of consensus building and sense-making in these two domains. Taking a cue from Latour, I have purposefully adopted a stance of symmetry in treating the epistemological regimes typically divided as Western and Other (Latour 1993). This challenges the universalism claimed by cultures of expertise such as those of the international aid world and its major institutions, a universalism that sees itself as having privileged access to nature, reality, and truth as opposed to the secondary knowledge of the nonexpert, one that is value-based, local, and thus sees the world not directly but through "social categories" (Latour 1993, 108). Adopting the symmetrical approach, I have instead placed the discursive orderings and their production via material practices of expert and nonexpert social worlds side by side rather than treating them as levels in a hierarchy of knowledge. From this perspective the Internet cafés of Accra do not generate merely an "indigenous knowledge" of the technology bearing only on local concerns. Rather the commentary of youth about the Internet should be understood as a contribution to the global debates over progress, technology, its relevance, and efficacy. The Internet café users, through their speech and activities in the cafés, offer an implicit critique of the assertions made at the WSIS.

There is a powerful association among Africa, poverty, and development in much of the affluent, industrialized world that makes it difficult to speak about the continent at all (outside of certain small specialist communities) without explicitly addressing issues of development. By difficult, I mean that to speak about the mundanity of everyday life in Africa, of TV-watching habits or fashion or football fandom lacks a discursive anchor and therefore seems to lack relevance. To analyze everyday technology use in Ghana in a way similar to that of researchers studying technology in advanced capitalist societies can seem frivolous and marginal in light of the widespread public representations, particularly within the media, of Africa as a continent in extreme crisis. How can one speak of computers when people are starving? This message of misplaced priorities is heard even among the techno elite who might otherwise be expected to champion technology for all. For example, Bill Gates commented in 2000, "The world's poorest two billion people desperately need healthcare, not laptops" (Helmore and McKie 2000, 5). The necessity of speaking about Africa through the lens of poverty and development has consequences for representations of Africans. Inhabitants of the continent are often portrayed two dimensionally as either barriers to development (e.g., corrupt politicians, warlords) or as development stakeholders in accordance with whether their expressions of agency are deemed legitimate or not by powerful international bodies such as the United Nations.

Over the years, as I have gradually amassed experience with the reception of this research by diverse audiences, the question of development has become a relentless companion despite my efforts to reframe the discussion. In presentations covering in detail the strategies of Internet scamming and the challenges of online representation and cross-cultural interaction, I have often been called on to conform my analysis to something largely irrelevant to the talk but according to dominant categories and concerns in development thinking. For example, I have been asked to indicate whether Ghanaian Internet users look up health information online or whether they find the Internet useful for business. I have occasionally been told by development practitioners that this depiction of how and why people use the Internet cafés in Accra is a powerful indication of the need to better educate new user populations about the broader (or perhaps proper) uses of this new technology. The concern with technology and its potential to facilitate betterment or progress of some sort is an apt concern but employing the lens of development seems to inevitably bring about a preemptive narrowing of the discussion. Thus I have delayed to this point in the book the direct examination of the Internet cafés of Accra as a phenomenon that might have anything at all to do with development. My resistance to this conceptual lens stems from the nature of development as a normative project such that an analysis along these lines slips too easily and too quickly into evaluation, toward deeming legitimate or illegitimate what is taking place in Accra's Internet cafés. It is far too early in the collective consideration of the Internet's true value to marginalized populations to fall back on such determinations. On the balance of my observations in Internet cafés in Accra versus in the unfolding proceedings of the WSIS conference, it seems that so far the relevant international organizations tasked with bringing about development (the United Nations and its agencies in particular) have far less of a handle on the efficacy and relevance of the Internet in places like Accra, Ghana, than the understanding young Internet café users have developed for themselves.

In her study of Tanzanian Internet cafés, Mercer describes as "African exceptionalism" a presumption that the Internet, although clearly a meaningful tool in the rest of the world for a wide range of communication, entertainment, and leisure functions, will instead find its proper place on the African continent as a tool oriented solely toward "productive purposes in the pursuit of market-based 'development'" (Mercer 2005,253). Definitions of development currently in circulation, although admirably expanding to include a multitude of dimensions, still suffer from a circumscription of meaning and value. Avoiding this black hole of thought is an effort to avoid the analytical impoverishment that such an insistence on a presupposed normative order imposed by conventional models of development.

In light of the powerful pull of the language of development, this chapter has been purposefully positioned after four chapters of substantive analysis situated in the Internet cafés, among young Ghanaian Internet users, and in the broader urban setting. The way the momentum of emergent technology practices in places such as Accra has been consistently overlooked or downplayed in venues such as the WSIS regional conference represents an enormous missed opportunity. This momentum might even be something aid organizations entering into the space of ICT access could usefully and efficiently support. The possibilities of thoughtful intervention (whether through alterations to policy, the design of new technologies, or pursuing programs to expand access) are numerous. However, to pursue this would require a far more fundamental and substantive reordering of how stakeholder and beneficiary groups are involved in development processes beyond the too superficial performance of inclusion that was on display at WSIS.

7 The Import of Secondhand Computers and the Dilemma of Electronic Waste

After the World Summit on the Information Society concluded in 2005, the attention of the national government and the business community in Ghana turned to a number of relevant concerns: the overtaxed electricity infrastructure, the influx of computers and other electronics as a burden on waste-handling systems, and the financial flows necessary for the business of enabling connectivity. Each of these issues illustrates a creeping awareness of the materiality of the Internet in Ghana countering a dominant rhetoric at WSIS that celebrated the transcendence of the material that would follow from joining the "information society." At WSIS, the instantaneous connection of the new network technologies was defined as a triumph over constraints of space and time.[1] The focus on information at the conference, as something formless and inexhaustible and to be made available to all of humankind, served as a savvy way to avoid all manner of intractable resource scarcity issues. This chapter works through some further issues of the political economy of the global Internet similar in scope to those raised in the previous chapter. However, by examining the hard infrastructural issues confronted in the actual build-out of Internet access in Ghana, this chapter serves as its counterpoint.

The primary substantive focus of the following sections is the emergent, ad hoc trade in secondhand computers imported from the United States or Europe by Ghanaian transnational family businesses. What is not considered in this chapter are some of the matters of Internet policy and network infrastructure that more regularly appear in academic analysis of technology and political regimes in the Global South. In Ghana there were some undoubtedly critical changes in national telecom policy in the early 1990s that paved the way for Internet access in general and the Internet cafés in particular. The subsequent founding of local Internet service providers (ISPs), the physical network and its management, and the particular interactions between Internet café owners and the ISPs are additional topics in this space that potentially might have been considered in these pages. The omission of this history is not meant to indicate its irrelevance; rather, these topics are especially well covered elsewhere (see Foster et al. 2004; Wilson 2006). By contrast, the computers that equip the Internet cafés in Accra and that are an essential conduit to the Internet have a perhaps unexpected story of arrival and circulation in Ghana that has not yet been told. Furthermore, examining the circulation of these machines helps to translate the material ambiguity of the Internet to the concreteness of material objects. Tracing their movement and the actors involved in their import, sale, and disposal touches on matters of national politics within Ghana, state-to-state relations, as well as global economic trends.

In the Internet cafés, property tags remaining on many of the Internet café computers identified schools, businesses, and government offices in the United States and Europe as the source for this equipment. The property tags I encountered and documented included ones for the New York Public Library, Anne Arundel Community College (in Maryland), the United States Environmental Protection Agency, St. Mary's College (in the United Kingdom), the University of Iowa, as well as tags written in Italian and Dutch. Such machines were likely considered obsolete and cleared out en masse as part of an effort to update IT facilities in these US and European institutions. The machines themselves were several development cycles older than the state of the art. It took seven years or more from the moment when a computer first appeared in the market to when it arrived at the Internet café.[2] The discovery of these tags in the course of fieldwork inspired this effort to trace and explain the circulation of these machines. It ultimately proved to be an illuminating and novel way to explore the political economy of the Internet in Ghana while maintaining a direct and materially grounded tie to the cafés themselves.

The import and resale of secondhand computers has played an underrecognized role in the success of Accra's Internet cafés as a model of public access to the Internet. Affordable and reliable equipment is a basic necessity any Internet café requires in order to sustain itself financially. Such equipment must withstand the often heavy use cycles, in some cases running for twenty-four hours solid in the setting of the Internet café. However, the upfront equipment investment must also be low enough that the occasional and unpredictable motherboard-destroying voltage spikes, periodic network and electricity outages, and repair costs can be absorbed by the business venture. These needs were met by Ghanaian traders through the development of this new distribution process, the trade in secondhand computers. This trade leveraged the short cycle of equipment obsolescence that is an established part of how Western (and especially US) consumer cultures regard electronics and high-tech commodities (Slade 2006).

In light of the ongoing discussion of marginalization that threads through this book, what is especially notable is the way these material requirements (unreliable infrastructure, twenty-four-hour use, and cost constraints) came to be addressed in Ghana by novel provisioning and distribution strategies rather than through product design. This may ultimately prove to be a larger trend whereby marginalized regions and their consumer cultures that are shut off from forums for influencing design proper (i.e., in the multinational corporations that initially define the specs, manufacturing process, and price points for these computers) instead find opportunities for agency and innovation in these later stages of a technology's life cycle. Ruth Cowan's influential work on the history of domestic cooking and heating technologies sets a precedent for considering distribution channels, manufacturing, and broader industry structures as critical to the success of a technology with consumers (Cowan 1987). In the current case this process of innovation was tied to the circumstances of a market largely invisible to or considered nonviable from the viewpoint of industry centers.

The broader ecosystem of distribution, repair, and disposal is necessary for the material support of Accra's Internet cafés. These cafés exist in the way they do because this supply of affordable machines is available. The nature of this supply furthermore had consequences for how users within the café engaged with the technology. Through this informal distribution channel, these computers arrived in Ghana lacking, at first, a strong framing narrative to indicate how users were expected to meaningfully or effectively employ them. The interpretive possibilities of the technology were not intercepted by dominant powers of the state or by foreign aid agencies. In an alternate scenario of state- or NGO-sponsored public access initiatives (such as the telecenters movement; see Roman and Colle 2002) efforts to shape Internet use may be carried out through training regiments and can also entail censoring and precluding certain uses that are deemed illegitimate by funders and organizers. Multinational corporations that produce computing technology (such as Dell, Apple, Intel, Microsoft, etc.) may also intervene and shape interpretations through advertising and marketing campaigns but their involvement in Accra was minimal as indicated by a lack of advertising, retail outlets, and sales offices there. Arriving through the back door of the global trade system, users were left to make sense of computers and the Internet drawing on the sorts of resources they had at hand. They did so specifically through small media formats (rumor), peer relationships, and other kinds of institutions (such as churches and related structures of belief) as the previous chapters have testified.

That the international circulation of secondhand computers appears, so far, to be beyond the interest and attention of the high-tech sector is likely due to the fact that this postpurchase trade represents no additional contribution to corporate revenues. Furthermore, the work of individuals and businesses that make up this industry in Ghana was carried out without coordination by any associative body, without a voice representing their interests to the government or to the media. The industry was not much more than an aggregation of traders and trading families. Many of its key players lived or traveled overseas for much of the year. However, this trade in secondhand electronics has recently become very visible in a very selective way through coverage in Western media outlets of the electronic waste (e-waste) problem in developing countries. Ghana, along with Nigeria and China, are the countries typically singled out for attention. The coverage so far has included an Emmy-nominated *Frontline* documentary broadcast on PBS in the United States on June 23, 2009, titled "Ghana: Digital Dumping Ground,"[3] an article in *National Geographic* titled "High-Tech Trash,"[4] and a glossy photo slideshow of the Agbogbloshie dump site in Accra on the Web site of the *New York Times*.[5] US-based environmental organizations, including Basel Action Network (an NGO dedicated to the e-waste issue) and Greenpeace, have fueled and shaped some of this attention.[6] In a strategic overhaul of the terminology for describing this import process, used computers have come to be referred to as "toxic cyber waste" (Puckett et al. 2005, 2) and those involved in the trade as operating in a "shadowy industry" that exploits regulatory "loopholes"[7] with little awareness of the health and ecological harm brought on their compatriots and their homeland.

That the trade in secondhand computers has, as of late, come to be framed through this new discourse of e-waste rather than the previously dominant metaphor of the digital divide has certain implications for shifting political will on some relevant issues of international trade. The trade in secondhand goods, in particular, has come under the scrutiny of a number of national governments in Africa. New legislation has been proposed in recent years to alter duty calculations and enact new standards that place penalties on this industry. In Ghana the overtaxed electricity grid is a key issue that comes into play. On a visit in 2007 I witnessed for myself the crisis of rolling blackouts as low water levels left the principle source of electricity generation, the Akosombo Dam, at well below capacity in a country developing an increasingly insatiable appetite for electricity. One tangible action is Ghana's recently passed ban on the import of secondhand refrigerators initially scheduled to go into effect in January 2011 but then delayed to January 2013. This was justified as a measure to stop a non-necessity that is especially prone to electricity overconsumption.[8] This measure impinges on the secondhand computer import business because importers often put refrigerators alongside televisions, stereos, and computers in the same shipping containers. Despite quite distinct functionality, both commodities flow through this same distribution channels, are often sold side by side in shops, and belong to a single commodity category from the perspective of importers and retailers.

For now there appears to be no serious discussion in Ghana to extend the ban from refrigerators to computers or other secondhand electronics or to impose or raise import duties on these goods. In relation to the rest of the continent, Ghana remains an especially easy country to import computers into. The duties that are officially listed in Ghana's most recent customs tariff schedules show 0 percent import duty calculated on computers.[9] This applies to CPU units, any kind of computer monitor (CRTs as well as LCDs), and computer printers. There is also 0 percent import duty on parts for any of these categories of equipment as well. It should be noted that no distinction is drawn between used and new machines within the Harmonized Commodity Description and Coding System (shortened to Harmonized System [HS]), a standard employed internationally as a system of names and numbers for classifying traded products. Customs officers do, however, determine a monetary value at the point of entry, ultimately drawing such a distinction. A standard 12.5 percent VAT tax is the only money collected by the government (apart from a 1 percent processing fee) on these high-tech commodities. By comparison other commonly imported electronics, in addition to the VAT tax, are charged an import duty of either 10 percent (for refrigerators, televisions) or 20 percent (for energy-sucking air-conditioning units) of their assessed value. For importers able to secure NGO status and who are bringing computers into the country to be donated to schools or other educational institutions, the VAT tax can be waived entirely.[10] By contrast, elsewhere on the continent, Uganda recently passed a total ban on the import of secondhand computers along with a range of other secondhand electronics, explicitly pointing to lack of waste-handling facilities as the reason, though the government has more recently been backtracking on the planned computer ban.[11] Kenya and Zambia have also recently begun considering such legislation.[12] This is a reversal from the kind of international political pressure around the turn of the millennium that called on low-GDP countries to reduce or eliminate import duties on computing equipment in the interest of facilitating IT literacy among their citizenry. It is perhaps also an indication of the success of those policies in bringing so many computers into the country that they are now coming to be recognized as a waste and disposal issue.

Strategies of Transnational Family Businesses in the Secondhand Electronics Trade

The import of secondhand electronics from the West to Ghana was a procedural innovation. It was sparked by the recognition of a growing supply of surplus, well-functioning machines that were seen as low value or even waste in the West, a supply then cultivated to be efficiently and profitably redirected to Ghana. Ghanaian migrants who kept a foot in both worlds played a critical role in facilitating this process through the formation and coordination of small family businesses that operated across national borders. There is an extensive literature on transnationalism that examines the kinds of interconnected social units that escape being determined and defined by the framework of a single nation-state (Ong and Nonini 1997; Levitt 2001; Smith and Guarnizo 2002; Waldinger and Fitzgerald 2004; Kivisto 2003). These social units do not necessarily constitute brand-new social forms but rather rework the more conventional orderings of families

(Bryceson and Vuorela 2002), political parties, social movements (Guarnizo, Portes, and Haller 2003), or businesses. The family businesses maintaining a link between Ghana and abroad epitomize this concept. Their success rested on navigating and patching together the policies and opportunities of two countries and mapping a course between divergent consumer cultures. Business owners sought profitability naturally but also ways to employ and provide an income for multiple family members

Although the number of Ghanaians migrating internationally has been steadily increasing since independence in 1957, there is a sense that especially desirable destination countries have become increasingly restrictive, barring all but the most affluent and well-connected Ghanaians. This perception likely follows in part from the fact that the population of Ghanaians desiring migration opportunities has become larger and more diverse (Goody and Groothues 1977; Peil 1995). Over the years, the desire among Ghanaians to migrate out of Ghana or to return has been punctuated by government changeovers and other key events. The severe economic decline and famine of the 1980s prompted large numbers of Ghanaians to leave. Certain events and the policy changes of foreign states have also shaped and shifted the flow of voluntary labor migrants. The policies of the British government enacted in the early 1970s ended the prioritization of immigrants from British Commonwealth countries such as Ghana. More recently, the attacks on the World Trade Center on September 11, 2001, and the subsequent economic decline in the United States have added further restrictions. There is also a perception in Ghana that Muslim travelers are particularly unwelcome. One response has been the turning of attention to a more diverse set of destination countries with South Africa, Saudi Arabia, China, and Malaysia among others emerging as alternatives that put up fewer barriers and restrictions on entry (Burrell 2008).

The ongoing scarcity of travel opportunities has contributed to the status and prestige of Ghanaians who have gone abroad and returned and who were known colloquially as *burgers* (or bɔgas as it is sometimes spelled in Akan/Twi orthography). This term is an abbreviation of *Hamburger* and its literal meaning is a reference to one who lives in Hamburg, Germany, a city that hosts a large community of Ghanaians. The term has come to be more broadly used to refer to Ghanaians returning from any Western country. The La Paz area (home of the Sky Harbour Internet café) was also an area where I was told that many bɔgas were operating secondhand electronics shops out of storefronts. As far as how family structure relates to the organization of these businesses it was generally a bɔga who was the primary owner and manager of the enterprise, the one who had invested his own capital as the initial start-up money for the venture. Siblings or nieces and nephews of the owner often did the in-country management of the business.

For example, one young shopkeeper Samuel who sold electronics that his uncle imported from the United Kingdom characterized the figure of the bɔga positively as one who is independent and who has experienced things. He added that such a person "knows business." The bɔga has people working for him rather than working for someone else. This was an enviable position that Samuel also aspired to. However, a more negative perception was that bɔgas could be "snobbish." This is how it was put by a scrap metal dealer named Ibrahim who works the other end of this commodity flow. Skepticism about whether the status of these bɔgas in Ghana was justified was evident in casual conversations and in representations in works of popular culture. For example, in a recent hip-life song titled, "bɔga, bɔga," the musician Sarkodie riffs at length on the way bɔgas claim a certain social superiority that is unwarranted given the reality of the very low status work such travelers often end up doing abroad. As the lyrics note, "Some bathe old ladies, wash underwear." Questioning the assumption of an automatic improvement in life circumstances by a move abroad, the song later suggests that if these bɔgas had stayed in Ghana they "would have completed school, would have been employed as a manager at Tigo." Tigo is one of the new mobile phone network providers in Ghana and employment there (as with the other providers) is employed as a sign of "making it" professionally. Nonetheless, by whatever means Ghanaians made their way while living abroad, the move abroad did prove in some cases to be a viable route for capital accumulation and for gaining certain advantages in trade as demonstrated by these importing businesses.

Data on computer imports from 2004 to 2009 obtained from the Customs, Excise, and Preventative Services of Ghana confirms that the trade in secondhand computers is, by and large, moving along the known labor outward migration routes of Ghanaians. In 2009 the top four countries from which computers were shipped were also the countries known to have especially large populations of Ghanaian expatriates: the United States, followed by the United Kingdom, then the Netherlands, and Germany (see table 7.1).[13]

Visiting the many small shops that sold secondhand electronics in the neighborhoods of La Paz and Newtown and talking to traders there confirmed not only that Ghanaians living overseas were facilitating and funding these imports, but also that most of the employees involved in such businesses were relatives. Generally the boss or general manager of the business was the one who did the work of traveling between Europe or the United States and Ghana collecting the items to be imported and other family members handled the import and customs process, worked as shopkeepers, and as technicians doing repairs on any broken items when shipments arrived in Ghana.

For these bɔgas, the business of importing secondhand computers and other electronics tended to be a more recent add-on to other ways of generating income. Some of these importers had first gone abroad as many as ten or twenty years prior in pursuit of work, money, and education. The circumstances of the trip were generally only sketchily outlined in interviews. In one instance, while visiting one shop my research assistant and I were scolded by the importer's mother for even raising the sensitive question of how her son had gotten his chance to migrate. The importer himself was out of the country at the time. Scornfully she muttered in Twi, 'what kind of question is that?' A move abroad could take place through

Table 7.1
Computers Imported into Ghana by Country Shipped From (2004–2009)

Year	2004	2005	2006	2007	2008	2009
Countries	Quantity					
United States	26,915	13,984	11,944	13,820	11,807	15,180
United Kingdom	34,113	3,494	6,264	4,863	7,961	4,784
Netherlands	11,518	3,323	3,819	2,189	2,379	2,555
Germany	4,705	1,204	959	1,446	3,260	2,356
UAE	8,048	803	543	554	1,418	1,818
Belgium	3,774	1,673	1,707	1,750	946	1,127
South Korea	1,684	63	235	672	2,836	1,028
Italy	1,456	375	350	530	738	1,024
Canada	4,054	1,275	4,751	2,216	1,147	1,022

nonofficial channels. At another shop, the young shopkeepers admitted that a "connection man," one who could falsify documents or pay off the proper authorities, was likely involved in their uncle's emigration. Marriage to a foreigner was another circumstance that had generated a travel opportunity for some computer importers. For those who had migrated decades prior and through unofficial pathways, the ability to go into the import-export business was an outcome of many years of working toward the financial accumulations and legal status that enabled freedom of movement between home and host countries. Mobility itself, because of its scarcity, became a lucrative resource that savvy Ghanaian businessmen and businesswomen were able to leverage.

The exact source of the imported computers obtained by Ghanaian importers was also difficult to get those working in these family businesses to specify. This was, after all, what gave the business a competitive edge. Some mentioned buying from middlemen at auction, others visited flea markets, some collected machines one by one from individual sellers. One said he sourced computers online from his home in Germany. It was possible to distinguish importers who had better business connections (and possibly better technical knowledge) based on visual inspection of the machines on display at their shops. The age, condition, brand, and homogeneity of the machines gave some indication. A stack of machines that were the same model and with the same specifications were likely to be the ones coming from institutions, were probably bought in bulk, and had a known history and interchangeable parts making the inevitable repairs easier to carry out (see figure 7.1). The less savvy traders had a more random assortment of computers, some off brand and perhaps even with key components removed. One importer named Kennedy who operated out of a shop in La Paz had a stack of about thirty or forty mismatched computers brought from the United Kingdom, all of which had the hard drives removed before he purchased them. He admitted to lacking any particular technical knowledge about computers and did not do repairs in-house. Computers were not the primary good he sold or where he made most of his money. He sold these as untested machines at a lower price shifting the risk over to his buyers. Because Kennedy sold the machines untested he had no way of knowing how many were irreparably broken (unless buyers came back to inform him or complain) but such an unselective importing process seems likely to contribute more in the short term

Figure 7.1
Shop of a top local computer importer; some machines still bear the property tags that reveal their source—the New York Public Library

to the electronic waste problem than the practices of more discriminating importers.

An example of a well-organized and coordinated family-run chain of shops was the one known as G.K. Asare Enterprises.[14] The two shopkeepers there, Freeman and Samuel, offered some details about the history of this business and how they coordinated with their uncle in his moves back and forth between Ghana and the United Kingdom. The signboard at the shop identified them as "Dealers in Electrical Appliences [*sic*]." It also visually depicted taxonomically each of the appliances they sold, a common practice in Ghanaian signage and a way of handling the wide-ranging literacy and language abilities of potential customers. From left to right the signboard showed a flat screen monitor, CPU unit, keyboard, speakers, and computer mouse, laptop, iron, microwave, stereo system, television, and DVD/VCD players. At the front of the shop a sturdy wooden table displayed CPU units stacked eight high and six across. They were a mixture of identical models and miscellaneous machines but generally ones I recognized as reputable brands: Compaq, Packard Bell, Fujitsu, and a few lesser-known ones. Each machine has a sticker on the front with three numbers written by hand: (1) the processor speed, (2) the amount of RAM, and (3) the size of the hard drive (e.g., P4 1.60/RAM 512/HDD 40 GB). A Pentium IV would be priced in the 170 to 250 cedi range (approximately $120 to $178) with the higher price including a monitor and peripherals as part of the package. These were the specifications that determined the price of the machine.

The day I visited the shop, the owner was away doing business in London. The shopkeepers were nephews of the owner and they worked as technicians and salesmen. They noted that their uncle went overseas generally for two or three months at a time. He had traveled to the United Kingdom for the first time about twenty years ago and began his import business (after many years of living abroad) bringing used vehicle engines into Ghana. Three years ago he turned to secondhand electronics because, as Samuel noted, he thought they would "move faster" (and thus repay the upfront investment quickly) unlike the engines, which could take months to sell. This family business had evolved over the years into a chain of shops. There were two other shops in addition to this one. Another one in the neighborhood was selling stereo equipment, secondhand refrigerators, and other domestic goods. A third shop was located in the city of Kumasi. The shopkeepers estimated that their uncle was filling and sending two forty-foot shipping containers approximately every three months.

Samuel and Freeman did work beyond simply minding the shop. Their uncle invested in his younger family members so that they could better contribute to the business. For example, Samuel was encouraged by his uncle to get training in computer hardware repair. He completed an A+ course at the local Wintech Professionals Training school (a tertiary school offering vocational training) to become a computer hardware technician. These skills he put directly to use in repairing and refurbishing the computers that were coming into the shop. Over time as the business developed, the uncle and his nephews had become more discriminating in their selection of goods to import as they came to know better what was in demand in the local market and what brands or models were easier to repair. The uncle called in periodically during his travels to get a reading on what would be worthwhile to import. The shopkeepers admitted that there were some early mistakes bringing in "those types they can't sell or when broken they can't repair" and leading to a certain amount of excess being passed off to the scrap metal dealers. So in the interest of the business the young shopkeepers "advised him not to bring [such items]." Samuel estimated that around 10 percent of the computers (meaning the CPU units) could not be salvaged and ended up being passed on to the scraps dealers, generally due to damage to the motherboard.[15] This fairly low number (to the extent that it can be trusted as accurate) was a result of their selection process ("my uncle is very good at selecting goods," said Samuel) and the fact that he and two other hired technicians worked hard to test, repair, and refurbish as many of the computers as they could. They were paid per machine and earned money only for what they were able to get working again, creating a clear incentive to salvage whatever could possibly be salvaged. This shop sold only tested goods that were confirmed to be working in contrast to the computers imported by Kennedy. This meant that when a buyer came to pick out such a computer, the salesmen would plug it in and boot it up for the buyer to verify its functioning, a service that was offered in lieu of a warranty.

The work undertaken by traders in the final stages of attracting customers and making the sale involved negotiating within a particular local consumer culture that, like any other, dealt in aspiration and the formulation of new wants and needs (Burke 1996). However, there were some particular issues raised by the fact that these goods had already been used by Western consumers. Although commodity purchases inevitably rest on the customer imagining a future improved by the acquisition of a good (Slater 1997), for these goods it had to do with imagining not just the commodity's future but also its past. Beliefs about the source and history of used electronics shaped perceptions of quality.[16] There was also a terminology specific to the category of secondhand goods. Used goods that were imported from abroad, including computers and electronics, were referred to locally as *home used*. The word *home*[17] referred to the machine's home, that is, its original consumer market. Some of the traders and consumers I spoke with also suggested that *home used* meant the machines were collected from people's homes. The term was meant to highlight that such goods were held to US or European consumer standards and specifications. The signification of *home used* was a positive one meant to indicate that the computer was affordable and also a high quality and reliable good.

This was the term traders often used when pitching their goods and bargaining with customers.

There was a tension between traders and consumers apparent in the way these groups divergently characterized and imagined the history of these secondhand commodities. In bargaining, given that the trader's goal was the highest possible price and the buyer's goal was the lowest, the description traders provided of these origins was necessarily treated with some suspicion. Another commonly used term, *aburokyire bola* (meaning literally *garbage from abroad*), was mentioned as a specific sort of insult one might lob at a trader, for instance, when bargaining had broken down or if a customer felt slighted in some way. It is worth noting that the outward facing term in English is the one with positive connotations whereas the insult is communicated via Twi. Consistent with Sarkodie's lyrics about the low status work bɔgas do abroad, this term went along with a general suspicion that secondhand goods being sold in Ghana were literally being scavenged from European or American dumps by Ghanaian importers. Generally, aburokyire bola referred specifically to the kinds of imported goods that were especially outdated and worn—faded and torn clothing, dirty-looking items, or items with missing pieces. These were the sort of things that were "well and extensively used" as a financier of one import business noted diplomatically. Importers took pains to correct the perception that they were acquiring goods through scavenging. Asking about aburokyire bola often produced defensive reactions among traders. One importer clarified that the items he imported he paid good money for. He asserted, "I don't call it bola [garbage] . . . if someone is selling it, you are going to use the money to buy. So how can you use money to go and buy bola?" The reasons that such seemingly good-quality items had been disposed of also demanded some sort of explanation. A major market in secondhand clothing in central Accra was humorously referred to as *obruni wawu,* which translates to *white man has died,* casting the market as a massive, year-round estate sale. A frequent concern raised by consumers was about disease transmission, particularly with clothing items or bedding. Joyce, for example, lamented the sale of used mattresses[18] in Ghana and noted, "some of the white people get sick, they will sleep on it and when the person dies, they will bring it here and some people can get infected if there was a person [who] died with serious disease. . . ." So although the term *home used* was used to promote the possession of such imported commodities as a way of establishing an equivalence between Ghanaian and Western consumers, by contrast aburokyire bola was employed to cast doubts, suggesting instead that consumers of this class of goods are reduced to the level of selling or using the (literal) garbage of more privileged others.

Electronic Waste Dumping and Further Dimensions of Marginality in Ghana

Having introduced the topic of garbage, the next step to consider in the circulation of secondhand computers is the journey of the defunct and broken computers after they exit the Internet cafés, homes, and offices of Accra. In the process of examining this end stage in the life cycle of machine circulation, another distinct youth population came to light, one that challenged a more routine understanding of the nature of human-machine interaction and highlighted broader forms of engaging the technology. This population was composed of the young men and occasionally children working as scrap collectors, scrap processors, and scrap traders. Computer units, computer monitors, and peripheral devices such as printers are increasingly a substantial part of what they trade in. Young men pulling wooden carts filled with metal scraps had become, by 2010, a visible presence in the streets of Accra (see figure 7.2). This is something that was not common in 2004 when this research began.[19] The interest in computers among these scrap collectors and dealers was not, however, in its capabilities as an intact and functioning machine. Rather they perceived the computer as an aggregation of more or less valuable constituent elements—copper, aluminum, plastic, and other materials. The work of these scrap dealers, their way of engaging with a computer, was radically alternate to the work undertaken at the machine interface by Internet café users. Their activities question the assumption that a technology's impact on a population stems from whether and how users interact with the intact and functioning machine.

Ibrahim worked as a scrap metal dealer in the La Paz area. Similar to most people involved in the scrap metal business in Accra, he and his family were from the northern region of Ghana. Unlike his fellow scrap collectors and traders, however, his family had been settled in Accra for a long time and his grandfather had established a family home there before

Figure 7.2
Young Dagomba men hauling televisions, a computer monitor, and other metal scraps in the La Paz area

Ghanaian independence in 1957. Ibrahim handled computers, monitors, and other defunct electronics as well as cars and car parts, iron sheets, iron bars and other building materials, domestic miscellanea, plastic goods, and more. He described himself as the "destroyer" for the area elaborating, "if something comes to me I have to destroy it. I doesn't repair. I destroy everything. . . . I used to cut cars into pieces and after I cut it, if they say I should repair it, I can't. I can use five minutes to cut a car [laughs] and if they give me ten days, I cannot repair it. So I have to call myself the destroyer. I always destroy." As his explanation indicates, Ibrahim defined his role according to a certain kind of power he possessed over material objects. Yet he joked in a self-deprecating way about the narrow specificity of the force he exerted. He could dismantle and obliterate with skill and efficiency but was totally incapable of reversing that process. He emphasized in a positive way the physicality of such work in relation to his tribe and regional affiliation asserting, "Those people who are working in scraps, they are the strongest men in Ghana . . . nobody have strong more than northerners in the whole Ghana." This work bore some similarities to other mechanical and machine-centered work, such as truck driving and auto repair, jobs that were disproportionately occupied by northerners and Muslims in Ghana.

Ibrahim was drawn to the work of scrap dealing by its low barrier to entry. In 2001 he had dropped out of carpentry school deciding that the burden that paying school fees placed on his family was too high. In scrap dealing, he found a job he was able to simply walk into without special training, degrees, or contacts. His mother gave him a modest sum to serve as his initial working capital, an amount of around $500 (5 million cedis in old currency circa 2001). This she acquired from the sale of a cow. He began straight away buying up metal scraps with this money to sell for a profit. He had arrived nine years later at an accumulation of around $5,000 in working capital, an amount he could continually reinvest in the business while still having enough money for daily needs and to provide financial support to his extended family. He had also found ways to augment his income by using his special position as a scraps trader to diversify his business endeavors. For example, two condemned cars he had acquired as scraps he was able to repair by enrolling assistance from his extensive social network of fitters (auto mechanics). Afterward he was able to get both vehicles operating as taxis. This generated a daily income for him that was enough to cover his basic needs. He was well connected with the many fitters in the area because they frequently had metal scraps to sell. Although he did not have the skills to do repairs himself, he certainly knew who could.

For scraps dealers such as Ibrahim, by far the most valuable component of a computer per pound was the copper wires that connected circuit boards, the power supply, and the ports inside the computer case. The copper wires had to first be removed from the plastic insulation they were encased within. Generally this was done by burning the wires to melt off the insulation. There was also a developing local market for the circuit boards. Many scrap dealers began collecting them on request from buyers. The buyers, as dealers noted, were typically Nigerians or Chinese. The Ghanaian scrap dealers who collected and sold these circuit boards did not actually know what was being done with these boards by the buyers but believed that they were being exported. The aluminum or iron frame of the computer case was also valuable. Scrap dealers traveled with the iron and aluminum scraps they had collected to the nearby port town of Tema, where they sold these scraps to the SteelWorks or Valco factories for recycling. Several scrap dealers said they believed the iron scraps were being recycled into iron rods to be used as security screens installed over windows and for other similar uses. The plastic shell and whatever miscellaneous parts were left after these extractions were generally dumped as is or burned to reduce volume. Computer monitors, specifically CRTs, generally had only one component for which there was a local market. This was the yoke that functioned in the CRT to direct the electrons that project an image onto the screen. This yoke was composed of a large copper coil. The plastic cases were generally burnt or dumped along with the glass screen. It was only the valueless elements that ended their travels right there in the nearest garbage pile, the rest went back into circulation, some to in-country factories to be reincorporated into new commodities and some went further afield re-entering global flows of raw materials.

The way Ibrahim tells his story, he depicts his entry into the scraps business as a choice he made willingly and that offered him an avenue for autonomy and self-advancement. Nonetheless the roles in this industry were not homogenous and there was a clear hierarchy of jobs. Below him there were the scrap collectors who had only their laboring bodies to employ—no working capital, no facilities, and few tools. By 2010 the wooden carts of these scrap metal collectors had become a common sight. They were a constant obstruction to traffic in the already jammed roads, were honked at indignantly by taxis as they moved at a pedestrian pace through city streets. For these collectors, their work entailed scouring the city (and beyond) for these valuable scraps and then physically, manually hauling them to a scrap yard or collection point.

Much of what is contained in these carts eventually ended up in Agbogbloshie, a scrap yard and dump site situated along the Korle Lagoon in central Accra. Agbogbloshie is a sprawling site, Accra's major center for the activity of processing, trading, and dumping metal scraps. Although the territory belonged historically to the Ga people who retain ownership of the land, the current inhabitants, by and large, were Dagomba. Conflict among different Dagomba clans and in particular a long-standing chieftancy issue (known as the *ya-na incident*) carried over into the social relations within the settlement. In August 2009 violence flared up between these two groups leading to a number of deaths but on more ordinary working days a visit to Agbogbloshie was more mundanely accompanied by the constant ringing of hammers thunking against metal and the acrid smell of burning plastic. Ibrahim's claim that the scrap metal industry involved the strongest of men in Ghana was convincingly demonstrated here. On one visit I saw several men leaning with every bit of muscle power they could muster to push forward an entire intact metal car frame balanced precariously on one of the standard wooden carts. The narrow but well-tread dirt trails threading through Agbogbloshie were traced and retraced all day, not only by scrap metal collectors, dealers, and buyers, but also by sellers offering food, water, air time cards for mobile phones, and a whole assortment of non-necessities, the typically extraordinary range of hawked goods. Just as the unfortunate circumstance of stalled traffic throughout the city attracted masses of hawkers, the revaluation and extraction of wealth from trash attracted similarly hardy microentrepreneurs. Along the main paved road that marks the boundary of Agbogbloshie there was a small mosque of cement block construction. A little further in, municipal government funding had established a football field that on one day hosted an informal game with teams who, lacking uniforms, played shirts and skins[20] while several men on the sidelines did a bit of betting on the outcome. Agbogbloshie supported a varied social world with provisions not only for industry and exchange, but also for religious practice and recreation.

Agbogbloshie could be subdivided into two areas, the scrap yard (as it is called) and, on the outskirts, along the banks of the Korle Lagoon, the *bola* (dump site). This area had been in flux for decades, settled only to be periodically cleared by the authorities. Reports of a proposed "clearing exercise" were once again appearing in local newspapers through 2009 and 2010. Agbogbloshie and the adjacent residential slum area provocatively named *Sodom and Gomorrah* were to be pulled down and vacated as part of the Korle Lagoon Restoration Project, though the threatened expulsion seemed to be indefinitely on hold.

At the furthest edges of Agbogbloshie, not in the scrap yard but in the bola, was where most of the apocalyptic images of sooty young men and billowing smoke that began appearing in the Western media were captured. This reporting drew attention to the issue of electronic waste export from the United States and Europe to developing countries. In the bola, the workers tended to be still younger than those in the scrap yard, teenagers rather than twenty-somethings. One fifteen-year-old I spoke with said he had migrated from the north by himself without family to accompany him or to receive him in Accra. The work these youth were doing was the simple processing of small amounts of copper scraps using self-fashioned tools to make money for their survival. They burnt the plastic insulation off copper wires, using chunks of insulation extracted from old refrigerators to focus and conserve the heat. The plastic insulation, for safety reasons, is fire resistant, making it difficult to burn. Young women with plastic bags of chilled water observed, sitting on overturned plastic computer monitor cases reappropriated as impromptu stools. They sold the sealed plastic bags of water they carried to these scraps processors who used them to suffocate the fires (see figure 7.3).

The work taking place in Agbogbloshie brought forward a whole new set of unanticipated risks stemming from this emerging computer trade and from the Internet café scene that was a source of demand for such

Figure 7.3
Extracting copper at the Agbogbloshie dump site in Accra

equipment. Local ways of processing through uncontained and open burning releases toxic substances harmful to the environment and to human health. The insulation around copper cables are typically made of brominated plastics (used for its flame retardant properties) and they release halogenated dioxins and furans on burning, substances that accumulate in the body once exposed and at high levels are known to be carcinogenic (Ladou and Lovegrove 2008). These chemicals are also known to suppress the immune system, interfere with hormone levels, and reduce reproductive capacity.[21] The CRT computer monitors (along with the television sets that were dumped alongside at the site) were especially troubling, containing only the small copper yoke that was of any value in the local market and large amounts of lead that was used to insulate against the radiation emitted by the internal components. Lead exposure can lead to damage to organs and to the central nervous system and reproductive system. Children are known to be especially sensitive.[22] Though the glass in CRTs was junk and wasn't purposefully processed or recycled, in extracting the copper yoke and getting rid of the valueless remnants it was inevitably smashed and left lying where all the lead, which had previously been suspended in the intact equipment, was free to seep into the soil or was pushed into the lagoon where it became a water contaminant. Among the scrap dealers, collectors, and processors in the bola and elsewhere I found that there was little knowledge or concern about any health issues related to material and chemical exposure.

These stories of the bɔga traders, the shopkeepers and technicians such as Samuel and Freeman, Ibrahim the successful scraps dealer, and the laboring scraps collectors and teenagers piecing together their survival from extracting and selling copper cables are all linked together by this flow of secondhand computers. Their diverging relationship to the computer also makes visible further divisions of center and periphery within Ghanaian society and helps to better situate the relative social circumstances of Internet café users in Ghana, the population that was the principle focus of previous chapters. A generational divide between youth and their elders and specifically the marginalization of the young by the old has been considered already. This chapter adds to this mix those populations inhabiting the economically depressed northern regions of Ghana as well as the general division between urban and rural populations. There is the marginalization, for example, most acutely felt by very recent migrants making their way within the urban center but without territorial claims, language skills, or expansive social networks. The various zongo neighborhoods in Accra, where large Muslim populations lived often characterized with a broad brush by outsiders, actually varied substantially as far as the duration and continuity of their existence. The neighborhoods of Nima and Mamobi already existed at Ghana's independence. A number of the young Internet users in Mamobi had been born and raised in Accra and were not confined as narrowly by small language communities. They spoke fluent Twi, the urban lingua franca, and often quite good English as well. They benefited from an education in urban schools and other resources whose quality tends to correlate with population centers. By contrast, the scraps collectors at Agbogbloshie typically were first-generation migrants, many arriving in the city only a few months or a year prior. The current Sodom and Gomorrah zongo adjacent to Agbogbloshie had formed only since the early 1990s or so. Many residents were part of a displaced population rather than voluntary migrants. Some number of these residents had fled localized conflict in the upper eastern region of Ghana. A marginalization related to the internal geography of the nation-state can be made concrete and visible to observers through the diverging ways of relating to particular commodities such as computers and other electronic goods.

Different social histories and political and economic resources may yield dramatically different ways of engaging with a technology. Scrap dealing was a particular mode of engaging an artifact (if not precisely a form of use in a conventional sense) that involved dividing the machine into homogenous elements, sorting out what was materially valuable toward recommodification. The gain was monetary in converting "junk" into something saleable. The work done on the machine through dismantling, burning, and sorting was a manipulation meant to add (marginal) value. Lacking the language abilities, literacy skills, money, and patience to engage the technology through the functionality of the human-machine interface, the scrap handlers brought an entirely different set of resources (the laboring body, physical strength) to bear on the artifact. Accounting for the scrap dealers in Agbogbloshie as another youth population shows how Internet access and use also implicates a necessary ecosystem of distribution, repair, and disposal. The dramatically different forms of engagement by the scrap metal dealers on the one hand and the Internet café users on the other ultimately map out very different distributions between the benefits and costs of engaging a technology.

Conclusion

The process of coming to terms with the, at times, harsh materiality of the Internet could potentially reshape the terrain of Internet access in Ghana in the future. Changing commodity flows and new regulations may follow from emerging domestic and foreign political pressures. This chapter has reconsidered the problem of electronic waste by looking holistically at the distribution channel that transports secondhand computers to Ghana and specifically the role played by Ghanaians in actively constructing this channel. So far, the formulation of the e-waste problem in Ghana by journalists and activists has not accounted for the distinctions Ghanaians themselves draw between what is reusable, what is valuable, and what is truly waste. Furthermore, the role played by Ghanaians as traders and the reasoning they attach to their involvement in this industry is largely excluded from such a depiction. The full set of issues at stake in this import process are much broader than this issue of waste handling alone. In addition to the way this particular commodity flow is harming the local environment and the health of those handling these machines, there are converse benefits to technical skill development and the employment of Ghanaian technicians who repair and refurbish nonworking machines locally and the technology access and utility that reused machines are providing to the general population in Internet cafés as well as in homes and offices.

From discussion with the computer importers themselves, the amount of true waste generated from a shipped container of secondhand or home used electronics seemed likely to vary substantially according to the different ways Ghanaian importers organized their businesses. This included their particular business skills as well as technical knowledge about the goods they were importing. Some importers were more discriminating and selective and better connected with supplies of quality secondhand machines than others. In terms of solutions, the emphasis of activist groups has been largely on the need to better regulate and enforce restrictions on the export of such used electronics from Western countries.[23] The need to set up proper waste-handling facilities in these e-waste destination countries has received much less emphasis by such groups. The center of gravity of this debate, in this way, remains situated in the United States and Europe. Yet the electronic waste issue is not a problem limited to secondhand machines but encompasses new equipment as well. All machines, after some period of time, will eventually become a waste-management issue so the in-country issue, as long as there is demand for and use of computers in Ghana, remains unresolved.

This chapter and the last bookend the six-year period during which the fieldwork for this book was undertaken (beginning with eight months in 2004–2005 and most recently seven weeks in 2010). They manage to capture the waxing and waning of interest and enthusiasm for the latest development solution—ICTs and specifically computers and the Internet. At what was possibly its climax following shortly after the dot-com boom, the global WSIS conference promoted the UN version of a utopian information society. The alarm over electronic waste is evidence that critical reconsideration is now well underway. The *Frontline* documentary on Ghana's e-waste problem makes a connection between this sense of promise followed by disillusionment noting, "when containers of old computers first began arriving in West Africa a few years ago, Ghanaians welcomed what they thought were donations to help bridge the digital divide" adding that this yielded only junk and new avenues for facilitating a criminal underworld.[24] Reviewing the history of development solutions, such a cycle appears all but inevitable. India's green revolution brought modern agricultural practices to boost crop yields but was met later with alarm over environmental degradation, overconsumption of certain farming inputs (water in particular) that led to social conflict, and the vulnerability of farmers to global markets and spiraling debt (Gupta 2000; Jasanoff 2002; Shiva 1991). The microfinance movement has recently been similarly reconsidered with concern over new bank lenders taking advantage of the poor.[25] Furthermore, academic work on borrower households found evidence of how loan programs sparked family conflicts (Rahman 1999). However, in the case of information and the Internet as a development solution, this cycle uniquely centered on renegotiating the materiality of what were previously presented as immaterial digital flows. The promise of a cost-free information revolution in this narrative arc came to be transformed instead into a literal river of waste.

8 Becoming Visible

Ghana, a small country on the Gulf of Guinea in West Africa, is the size of the US state of Oregon. Its entire population is only double that of New York City. Yet what is unfolding there, I argue, matters to the future of the Internet. By exploring the social world of youth who inhabit Accra's Internet cafés, on their turf and in their words, I have sought to contribute to a richer understanding of how digital technologies might be desirable and useful in the world's peripheries. Much of the early conversation about the Internet centered on its liberatory possibilities. Its material properties were linked with certain social ideals—equality, openness, and freedom. The increasingly diverse user populations now online—as represented, for example, in the cross-section of humanity present in Yahoo! chat rooms—offer a more rigorous check on whether these ideals have actually come into being, whether the material terms of online participation indeed communicate and support such ideals, and whether such ideals are shared or ultimately adopted by these newcomers.

When young Ghanaians went online, they did not join the digital age unproblematically. Instead they occupied an unaccounted for in-between space, neither simply connected (and conforming to a Castellian network logic) nor disconnected and wholly excluded. The old dichotomies, it is clear, no longer suffice to frame issues of inequality and inclusion in global networks as the network expands to encompass more and more users from diverse geographies. Scholarship must begin to contend with this new marginality, one shaped by connectivity under various asymmetries of material connection and differences in representational power online.

The fate of culture in an era of networks has long been of interest to scholars. The thinking that closely identifies culture with tradition and locality typically has pointed with concern to processes of cultural homogenization as the likely outcome of the bleeding boundaries of global interconnection (Ebo 2001; Herman and McChesney 1997; Mattelart 1985). The suggestion here is that cultural contact generally yields to one culture over another, yet how aspects of culture come into being in the first place or by what processes culture is changed is unclear. Manuel Castells's claims about the cultural dimensions of the digital age, though more nuanced, nonetheless similarly rest on a notion of culture that in light of the present case now appears too neatly apportioned and bounded. He partitions a global culture that is constituted in and through the network from the multiplicity of enduring, revitalized, or radicalized regional cultures (Castells 1997). Of this global culture, he argues that founding communities (of entrepreneurs, hackers, and idealists particularly concentrated in the United States) have shaped an ethos of online spaces, their configurations and possibilities, producing something Castells calls "a culture of real virtuality" (Castells 1996, 355). Castells describes resistance and the promotion of cultural alternatives in regionally anchored social movements. The Zapatista movement of Chiapas State in Mexico is one of his recurring examples. At times these reactive movements manage to employ the network itself toward their cause, yet in Castells's formulation they do so without apparently deforming either the network or their causes. An absolute differentiation of global and local is in this way retained.

Castells depicts an oppositional cultural response as something orderly and intentional. In his examples it is an explicit articulation of alternate values and a willful attempt to preserve them. Yet the way Ghanaian youth interpreted the new material possibilities of the Internet could not be characterized as conscious resistance (with the exception of certain strategies among scammers) nor could it be adequately characterized as a kind of social movement. These youth operated in part through inculcated dispositions and desires (their habitus), which came to shape their engagements online (Bourdieu 1977). The unconscious ways that members, once fully socialized, inhabit a cultural position is what is notably omitted from Castells's consideration. Yet this is what was especially apparent in the confused cross-cultural encounters online between young Ghanaians and foreigners as a result of the unforeseen differences in baseline expectations they held about social roles and interactional norms.

Writing early in the Internet's history, Castells saw a trend toward uniformity in global systems that were "increasingly speaking a universal digital language" (Castells 1996, 2). However, with the more global makeup of the Internet's population of users the system now seems likely to take the opposite course. As divergently socialized groups go online they make over the cultural topography of the Internet unevenly and heterogeneously. When Castells speaks of a "universal digital language" the word *language* misleads. The Internet's global interoperability rests on what is more aptly labeled a technical protocol, not the same thing as a human language. Although the technical mechanism may indeed be universal, above this layer, in the actual exchange of human language, this is far from the case. Online balkanization today often marks the boundaries of language communities, for example, the distinct Portuguese, Chinese, and English social networking applications (Orkut, Xiaonei, and Facebook, respectively) that demonstrate continuing separation within a globally connected system. The struggle for authenticity and representation among young Ghanaians online, as recounted in this book, shows how formal language differences are not the only barriers to mutual understanding. Within the English language–speaking Internet, encounters that cross cultural distances also easily yield confusion and incoherence in the clash between meaning systems.

The following sections delve into some implications of changes taking place in Ghana's urban Internet cafés in recent years. There is a growing visibility of Ghanaian Internet users (particularly scammers) within Ghanaian society as well as to more distant users, site builders, and administrators operating from network centers. An evolution in the social practices of the Internet in Ghana during the duration of this project (2004–2010) now intersects in some interesting ways with more geographically distant trends, such as the recent network neutrality debate in the United States.

The Rise of Sakawa

When I left the field in 2005, I left more or less completely, returning for only a brief three-week visit in 2007 to pursue unrelated research. In 2010 when I returned with the intention to catch up on changes to the Internet café scene there were some very sharp and apparent contrasts. With my carefully studied fieldnotes circa 2004–2005 cemented in my mind during the ensuing years of analytical work, I saw the present era through the lens of this recent history. I set about re-interviewing eleven of my key informants and found greatly reduced Internet use for some and intensified use for others. New forms of access through GSM modems allowed home Internet access and several had shifted from the cafés to home use. Many Ghanaians were now on Facebook. All but one of the Internet cafés that were key sites for my research remained open and operational[1] suggesting that this trend was not a fleeting or financially unsustainable one.

Across the range of Internet practices, scamming had changed the most and had the greatest impact in refiguring the Internet café scene. It proved to be an especially dynamic and increasingly effective activity and had generated a distinct subculture—that of the sakawa boys—as will be detailed in this section. In the case of Gabby, the young scammer whose activities and views of the Internet were considered several times in previous chapters, he eventually hit on a successful scam strategy and gained somewhere in the range of a couple of thousand dollars. When I met with him once again his living conditions were much improved. He was renting a better room and had nice furnishings and clothes, his own computer, and a home Internet connection. He owned a couple of trucks he rented out to earn a little income. He used these funds to produce his girlfriend's gospel music CD (this girlfriend was now his ex) though he noted that after all of this he was broke once again.

As with any activity considered deviant, where both the perpetrators and victims are reluctant to admit involvement, it's difficult to say whether or how much Internet scamming in Ghana is truly growing in terms of the number of scammers involved or money gained. In more recent years, however, scamming activities have begun to generate growing public alarm among authorities in Ghana—parents, politicians, preachers, teachers, and business leaders—and this new attention is a very clear and significant change. Internet scamming became widely referred to in public media outlets in Ghana as *sakawa*. Back in 2005, *sakawa* and *419* were terms quietly disclosed in conversations with Internet users but did not circulate much beyond the scammers themselves. Much of what was known circulated via secondhand stories—small media (as in the rumors examined in chapter 4) rather than mass media. This perspective was best summed up by one young woman, a regular visitor to the Internet cafés in Mamobi, who on the matter of Internet scamming noted, "I don't know of anyone who does that, but I hear people do it." *Sakawa* was a term used only in the zongo neighborhood of Mamobi. By 2010, it seemed that certain known Internet cafés in specific neighborhoods (Mamobi in particular as well as Nima, Newtown, Alajo, and Nii Boi Town) had been wholly overtaken by Internet scamming. A new practice I observed at cafés identified with scamming was the use of mirrored, adhesive window coverings or curtains to prevent prying eyes from seeing inside. Certain cafés, I was told, came to function as de facto private clubs where customers who were not known to the operators or regular users of the café were unwelcome. Internet café users and operators were said to literally lock themselves inside as they worked into the late night hours.

These shifts in the Internet café culture raise doubts about whether the opportunities for exploration and virtual cosmopolitanism, pen-pal collecting, the mischief of youth in groups, and the escape from disciplining elders enjoyed by young urban Ghanaians only a few years ago will still exist into the future. With a more public association between the Internet cafés and criminal activities, the scripting of these spaces and the technology within are becoming more entrenched and inflexible. This may pose a barrier to prospective new users and to some of the broader uses of the Internet cafés if these spaces come to be dominated physically or in the public imagination by scamming practices. Conversely, perhaps, the particular form of power now publicly attributed to these spaces and activities will draw more individuals interested in big cash gains and scamming activities.

Formerly regular Internet café users attributed their less-frequent visits to these spaces to what they perceived as incivility and a lack of trust online. Kwadjo laid out his bottom line: "I want to see that this person is a genuine person that I'm dealing with." Joyce pointed specifically to the chilling effect of the scammers on nonscamming uses of the Internet noting, "because of the 419 people, somebody can bring your picture, you will see and you will think oh, I like this lady, I want to be her friend, but when you go you are talking to a man. That's why I've stopped going to the Internet." These comments also reflect the perception among former users that it was not just the domination of scammers in the cafés themselves or the new public attention within Ghana but the way this changed perceptions of the online, introducing a new skepticism about digital self-representation that made encounters more paranoid and fraught.

As for Joyce, whose marriage to George was described in the first pages of this book, I found on my return visit that their marriage continued, though they had spent very little time together in the intervening years. Yet, she too was benefiting from a better material standard of living. Like Gabby she had a nicer room, furnished with several new (not secondhand) kitchen appliances, and recently had acquired a system engineering certificate from a reputable IT training school, with George's ongoing financial support. Moving Joyce to Canada proved more difficult than they had anticipated. The couple had not yet managed to secure a travel visa, though Joyce remained optimistic that it would happen in time. She no longer saw any appeal in visiting the Internet café, for the reasons described previously, but she also noted that she'd gotten from it what she had most hoped for in George. Her communications with George took place by phone as new, more affordable international calling had become available. For many former Internet café regulars, changes in life circumstances—a new job, schooling, or in Kwadjo's case, the birth of his daughter, occupied the time they had once spent staving off loneliness or boredom by chatting online in these public spaces.

Outside of the Internet cafés themselves, Internet scamming had generated a visible and ostentatious material culture. It was now easy to find convincing evidence that such strategies were yielding significant material gains. The style signifiers were consistent and widely known and included certain types of clothing and hairstyles. Sakawa boys were widely recognizable urban figures, wearing shades, gold chains, new clothes purchased from local boutiques (rather than the secondhand clothing markets), and Timberland shoes, a style drawing influence from the American hip-hop subculture. They drove a newer model Toyota Corolla, Camry, or Matrix; a Dodge Charger; or a Chrysler with an unregistered license plate indicating that the car was newly imported into the country. It was also not hard to find the Sakawa boys hanging out at certain night spots. At one spot, a weekly Karaoke night drew such crowds. The night I visited, aspiring rappers took turns freestyling over beats. Later a popular Nigerian gospel song came on to the delight of the crowd who began to dance at their seats. Meanwhile along the street young men in cars cruised up and down to show off their vehicles. One car full of young men overzealously steered off course becoming lodged in the concrete sewer ditch, wheels spinning impotently. A gamelike and surreal feeling was evoked in this nighttime parallel world of maximized consumption and meaningless, throwaway money.[2]

References to sakawa in popular culture had also become frequent by 2010 as the term found its way into the urban vernacular. For example, one day I came across a street preacher using a loudspeaker to exhort youth to avoid being persuaded by their peers who do sakawa to join in on such activities. The term also began appearing in newspaper headlines, sometimes in small blurbs about arrests or court cases, at other times in lurid front-page stories. For example, a short piece in the most widely circulating newspaper, the state-owned *Daily Graphic* documented an arrest with the headline, "Man, 20, Nabbed over 'Sakawa'" (*The Daily Graphic* July 10, 2010). A local private newspaper had a front-page article titled, "Sakawa Murder in Nima," which documented the case of one Internet scammer whom, witnesses believed, had been beaten to death by his Rasta[3] co-conspirators in a scam, presumably in a conflict over money (*The Daily Guide* July 20, 2010). Nigerian and Ghanaian movies have picked up the theme with titles such as *Sakawa 419* (parts 1–4)[4] and *The Dons of Sakawa*. In addition, street vendors who sold locally designed, color-printed posters of recent events and topics of interest began sometime around 2008 to carry ones titled "Stop Armed Robbery & Sakawa" and "Sakawa Money" alongside the more typical posters commemorating important foreign visitors (US President Barack Obama, Queen Elizabeth II of England), sports teams (local and European soccer), and local political figures.[5] The term *sakawa* also came to be part of a more general moralizing discourse on legitimate and illegitimate avenues to wealth and the distribution of funds. For example, after the Minister of Finance, Dr. Kwabena Duffor, delivered a statement about the national budget and the fiscal policies of the ruling National Democratic Congress (NDC) back in March 2009, members of the political opposition publicly denounced it as a "sakawa budget."[6] The word entered into regular usage as a broader reference, well beyond the Internet, to describe any practice at all that was considered fraudulent or dishonest.

Reflecting the increasing visibility and consequent anxiety over these activities, sakawa also came to be associated in emerging public understanding with spiritualist practices, the mechanisms of juju and "blood money." One circulating narrative about practices of Internet scamming was that these youth enrolled dark forces by participating in certain rituals as a way to render effective their online activities. Blood money was arranged by visiting an unscrupulous fetish priest or a mallam who was willing to direct spiritual forces toward such destructive aims. Generally this involved the sacrifice of a close family member, such as the scammer's mother, wife, brother, or child, who would die suddenly and inexplicably before gains from the scam were realized. The zero-sum game that such rapid and ill-gotten wealth was thought to be subject to led to the belief that scammers who made such deals would also live shortened lives. A policeman who observed many presumed sakawa boys while working traffic stops told two stories of car accident scenes he had been called to that involved fatalities. The death of a young man in this circumstance was a compelling indication of the victim's involvement not just in scamming, but also specifically his trade in blood money.

The flourishing of these tales confirms that the attention paid to rumor in chapter 4 was not a tangential concern but that such informally circulating messages are an important mode of grappling with the Internet in urban Ghana. These sakawa tales offer a new round of rumor, ultimately becoming more formalized in popular culture and the mass media. The blood money stories in particular, on one level, reflect the distanced mystification of the technology among nonusers, their uncertainty about how it operates to yield such wealth.[7] The scammers I spoke with denied using these techniques,[8] referring instead to their skills of language and persuasion and the evolving format of the scam as responsible for increased efficacy. How prevalently spiritualist practices are employed to enhance Internet scamming remains quite unclear. There was indeed the tone of a moral panic to these stories (Cohen 1972), specifically in the way that they served to amplify and personalize concern about these practices within the broader urban society in Ghana.

The new sakawa narratives extended the moral scripting of the technology into the mass media and reworked it in new ways. Previously such narratives appeared only in the justification of scamming accomplished through small media formats such as rumors. Societal fears and moral approbation were clearly mapped out in this new configuration of relationships and forces, specifically in the way the victimization of a foreign stranger via the Internet was relocated to the scammer's intimates and to the scammer himself. A conclusion one may draw is that despite the absence of much effective legal enforcement by the Ghanaian police and court system, despite circulating representations of the wealth and greed of Western scam victims, such ill-gotten gains from the Internet were far from a socially accepted practice in Ghanaian society.

On the Neutrality of the Network

The new visibility of Internet practices and their reworking in sakawa narratives circulating in urban Ghana showed an informal action against scammers via censure and social pressure rather than through the law or through technological reconfigurations. A similar visibility of Ghanaian users within the broader global network, where they are coming to be viewed principally as a security threat, has, however, produced new technological and material barriers that concerningly compromise Ghanaian Internet users' open access to the Internet. Rather than putting Ghanaians' activities in check on moral grounds, this response is a product principally of a trend toward the commercialization of the Internet. Such a trend puts the issues of the network's openness and its neutrality in a new light.

The current debate over network neutrality in the United States has once again enlivened discussion about a set of values and ideals tied to the Internet and whether they can or ought to be preserved. The convergence of television and the Internet now underway and the shifting of Internet data transmission from telephone lines to cable infrastructure has presented a challenge to bandwidth management and revenue generation. This meeting of media models, the closed broadcast model of the TV world versus the open and decentralized model of data flow on the Internet, must somehow be managed though the interests of content providers, network infrastructure maintainers, access providers, and users are in some ways at odds. The key issue under consideration in the network neutrality debate is what forms of revenue generation should be permissible for Internet service provider companies. ISPs such as Comcast and Verizon in the United States have some clear reasons in the future to pursue privileged or tiered systems of payment related to bandwidth demands and data routing, possibly downgrading or blocking certain forms of content or content from certain sources. This scenario evokes as a dystopian extreme, a corporate-controlled mass media overtaking the byways of the Internet, shutting out smaller or weaker or simply undesirable content or application providers as determined by commercial interests. At the level of policy, there is a discussion underway around legislating requirements that will ensure the continuation of the agnostic treatment of bits and bytes as they are currently passed along the Internet infrastructure, just as the telephone network came to be held to this neutrality through regulation in the AT&T break up in the United States, which went into effect in 1984 (Lessig 2001).

A stance favoring network neutrality generally upholds as one of its principles that users ought to be able to access any and all legal content and services without discrimination. Advocates argue that the neutrality of broadband infrastructure is necessary to maintain the valuable role of the Internet for supporting open political debate and freedom of speech for individuals and society. Considering the consequences of a lack of neutrality for markets and economies, some argue that this is likely to impinge on innovation and competition in the Internet industry (Peha 2007). The US Federal Communication Commission (FCC) and companies such as Google have released numerous policy documents[9] on the issue. The reference points they contain are by and large transitions in US technology, information, and media industries. The dilemmas of network neutrality appear to be, so far, considered exclusively in the context of the US legal system and, it seems, primarily with consideration for US consumer markets. How changes resulting from new regulations (or the decision not to regulate) will scale out to the global Internet and particularly to emerging markets, developing regions, and other peripheral populations of users is unclear.

Barriers already exist on the network that selectively exclude populations from accessing significant portions of the global network. Yet some examples of the Internet's broader non-neutrality are very high profile and others seem not to be discussed at all. Arguably the most widely known example of this is the filtering and censoring of Internet content by authoritarian regimes to prevent access by their citizenry, such as the case of China's "Great Firewall" (Kalathil and Boas 2001; Deibert et al. 2008). By comparison, a less-acknowledged practice of blocking countries on the periphery of the global economy takes place via IP address detection with the intent to prevent users in those locations from accessing certain contents and Web services. Typically this is done as a security measure by network administrators and site owners who, recognizing that a recurring security threat is coming from a locatable region, respond by blocking all traffic from that region. The West African countries of Ghana and Nigeria as well as many former Eastern Bloc countries are common targets for exclusion. This is not precisely a violation of network neutrality (as it is currently defined) because the barriers are configured at network end points rather than in the infrastructural routing algorithms and configurations. However, it raises some larger issues that are part of the network neutrality debate, specifically about open access to the Internet and the entitlements of users. Thus far it has been considered the prerogative of such providers to determine who to permit entry to the resources they provide. Password-protected, invite-only online communities can be traced back to the early days of the public Internet. In some ways this practice is no different.

The experience of coming across blocked sites while in Ghana is what brought this issue to my attention. For example, trying to visit westelm.com (the e-commerce site of a furniture store) I was faced with a brief message on a Web page informing me of the inaccessibility of this site from my location. I was not merely blocked from shopping functionality, but also from browsing any of the contents of the site. In another incident around the same time, after logging into my account on amazon.com, I received an email that my password had been changed automatically. In this email I was warned that I may have been the victim of a phishing scam presumably because my IP address on login resolved to a location in Ghana. This was not as much a wall as a detour put up on the Internet but it shows a kind of encoded suspicion Internet users in Ghana face when they go online. Additionally, the travel Web sites Expedia, Travelocity, and Orbitz do not permit Accra, Ghana, as a travel destination for online airline ticket bookings. Two of these sites, in fact, recently removed Accra from its listings though it had previously been available as a destination for ticket bookings. This was also presumably done for security reasons. Because Webmasters and network administrators don't generally make their security configurations transparent, many of these erected barriers are only apparent when navigating the Internet from within Ghana. Thus from the perspective of network centers it is a largely invisible problem, one left to this population of users in the periphery to push back against and make known in the centers.

The encoded barriers Ghanaian Internet users faced online fell into two categories: failure to include and purposeful exclusions. The most apparent and perhaps significant example of failure to include was the lack of accommodation on the vast majority of e-commerce sites for users from cash-based economies. Such services generally continue to serve only those users with access to international electronic credit networks. There was awareness and resentment of this particular exclusion in Ghana. The membership fee at the dating site Match.com, for example, was one example regularly offered by young Ghanaians. This was exactly the sort of service they were greatly interested in and willing to pay for and yet there was no way to do so. Credit card fraudsters sometimes pointed to the absence of legitimate payment mechanisms as a justification for their practices. The primary alternative online payment mechanism, Paypal, to date does not permit money transfers to and from Ghana or Nigeria (though it does for many other African countries) in light of security issues. On this specific matter of electronic payment systems, the Ghanaian expat blogger Koranteng Ofosu-Amaah of the blog *Koranteng's Toli* in a post on inaccessibility and network exclusions thoughtfully lamented, "if we take ecommerce as one component of modern global citizenship then we are illegal aliens of sorts, and our participation is marginal at best."[10] Such a commentary shows specifically how a sense of global marginalization endures and may in fact be felt more acutely as Ghanaians go online and are thwarted when they attempt to seize what are for most users considered to be the entitlements of online membership.

More purposeful exclusion often takes the form of IP address detection and blocking at the country level. E-commerce sites often do this when problems of credit card fraud seem to frequently originate in Nigeria and Ghana. The public Internet chatter[11] among Webmasters and network administrators often reflected little concern about or resistance to scaling a geographically locatable security threat to country-level or even possibly continent-wide traffic blocking. "I am so fed up with these darn African fraudsters," one poster at WebmasterWorld.com fumed, "is it possible to block african traffic by IP?"[12] On another site a poster goes a step further to suggest wholesale network exclusion, "Maybe we could just disconnect those countries from the Internet until they get their scam artists under control."[13] A site promoting tips and code samples for Linux administration points to revenue concerns as justification for banning: "I admin [an ecommerce] website and a lot of bogus traffic comes from countries that do not offer much in commercial value."[14] The responses to these queries expressed some skepticism about the effectiveness of such configurations but not the ethical or ideological grounds of this practice except for one commenter who suggested this move was "highly questionable . . . to broadly block users from entire countries from accessing e-commerce sites is unethical" then outed himself as "an American engineer working in the oil fields of Nigeria." He added that he "has to rely on Internet access to [*sic*] banking, insurance, educational and technical sites just as you ALL do. If you block my access because I'm on a Nigerian IP address, I'm screwed." The reaction was less than sympathetic, with one commenter suggesting that "for Americans living outside the USA move back home or buy a subscription to a proxy service." Such statements reflect a weak and contingent commitment to ideals of openness and accessibility of the Internet among some of the rank and file who maintain the Internet's many, many nodes.

Such security configurations are also increasingly employed at online dating sites that (unlike these e-commerce sites) have in the past been available to and very popular among Ghanaian Internet café users, among scammers and nonscammers alike. The blocking of access to Ghanaians, Nigerians, and others is troubling in some additional ways. The come-and-marry scammers who frequent dating sites have taken to reconfiguring their IP address in the Web browser in an attempt to reroute through a US-based proxy to evade such barriers. This practice of what is called *IP address spoofing* has become one of the essential tricks of their trade. Scammers were on a constant hunt for new dating Web sites, whether for straight or gay singles, for specific ethnicities, for overweight and older people, as they found doors rapidly closing against them on these sites. It was a race against time to gain contacts before a profile was deleted and a user account banished. Of course, the result was also that the same blocks and barriers were put up against earnest Ghanaian daters seeking a genuine person overseas for companionship, love, support, and marriage. We can recall the example of Joyce, who found her Canadian husband on plentyoffish.com to highlight the diminishing possibilities that come with such increased vigilance on the part of dating sites. Joyce also had her account on plentyoffish.com, the dating Web site she used, shut down after a time way back in 2005. Plentyoffish.com now notes publically that they block all traffic from Africa, Romania, Turkey, India, and Russia "like every other major site."[15]

Speaking with scammers and other Internet users in Accra, they perceived a process of escalation from the banning and blocking functionality available to individual users in software such as Yahoo! chat, Skype, and Facebook (as considered in chapter 3) to a more absolute, preemptive exclusion by network administrators encoded in site configurations. On the dating Web sites, one can see how such Web services are being configured to supersede the human interpretive capabilities of their users. In some cases these sites now make broad and exclusionary social judgments on the behalf of their preferred user base. The shift toward IP address blocking and other administrative filtering makes exclusions systematic, total, and materially concretized, no longer a matter of a case-by-case judgment of individual users at the content level but rather programmed into the non-negotiable realm of service configuration by Webmasters and network administrators. This meant that when Ghanaians encountered these barriers they no longer had the opportunity to persuade and win the attention and trust of the foreigners they encountered.

Castells speaks about the way "global networks of instrumental exchanges selectively switch on and off individuals, groups, regions, and even countries according to their relevance in fulfilling the goals processed in the network, in a relentless flow of strategic decisions" (Castells 1996, 3). There is something of this dynamic in these cases of withdrawn access. However, I wish to revisit and revise his assertion about the impersonal network logic of this process to point to the decision-making role of real human actors with their own baggage of perceptions and prejudices. Close up one can see how users and network administrators who occupy more privileged positions at network centers draw on a mapping of the world, of regional populations and their characteristics. This mapping becomes increasingly coarse and heavily mediated as it moves further away from direct experience. Parochial attitudes thus come to shape network infrastructures. Legitimate security problems posed by the regional scamming subcultures are handled through algorithmic formulations channeling network traffic in ways that are effective for security interests but demonstrably overreach their intended targets.

Finally, I wish to engage here in a bit of cautious speculation befitting a conclusion by considering how national policy discussions in the United States might unfold into global consequences. If network neutrality falls in the United States, could this pave the way for the sorts of exclusions that already exist at the site and application level (at these various network end nodes), moving more insidiously into the network routing infrastructure? What prevents an ISP from blocking network traffic from Nigeria or Ghana altogether to protect its customers? If revenue generation is accepted as a legitimate concern for shaping the prioritization or even blocking of network traffic, is there any obligation to serve the population of users, particularly those largely peripheral populations who use this infrastructure but who are not now and are not likely to become paying customers? The potential is present (and may in fact already be underway) for a process of *abjection* of particular real-world geographies from cyberspace. The anthropologist James Ferguson first employed this term to capture the way Zambia's post-independence decline was experienced by Zambians as a process of "being pushed back across a boundary that they had been led to believe they might successfully cross" (Ferguson 1999, 236). Specifically this meant crossing over from colonial paternalism and secondary status to membership in a "new world society" (Ferguson 1999, 234). As an example, Ferguson describes another sort of network, that of the international airline routes that once traced routes between Zambia and major and not-so-major world cities until the country's economic crisis when "the European carriers one by one dropped Zambia from their routes" (Ferguson 1999, 235). Ferguson's example of air transport as well as Ghana's Internet security crisis challenge notions of globalization as a teleological process of irreversible and increasing interconnectedness.

In his *Declaration of the Independence of Cyberspace,* John Perry Barlow wrote of cyberspace as its own self-organizing world, without need for government and beyond the sovereignty of any nation-state (Barlow 1996). The declaration addressing "Governments of the Industrial World" and written as the voice of citizens of cyberspace asks simply that their virtual society be left alone. Yet new trends toward corporate ownership of Internet infrastructure suggests there is a very real threat to certain early ideals of the Internet, not just from potential government impositions but also from the commercialization of the Internet, as analysts have more recently argued (Lessig 2006). Furthermore, there are the previously noted security issues that scamming activities present and that are so far unchecked by relevant policing institutions. The role played by real-world governments ought to be reconsidered in light of these new realities. A too-weak police and court system in Ghana has left scammers to pursue their gains largely without resistance. Scammers did not fear the local police (whom they intimated were receiving a cut of the gains made from scamming), though family pressures and societal stigma did compel some young scammers to quit. Matters of jurisdiction further prevent legal action by the countries where victims reside who cannot reach these foreign perpetrators. Although Barlow asserted that in cyberspace "we are creating a world that all may enter without privilege or prejudice accorded by race, economic power, military force, or station of birth," this recent history of the Internet suggests that governments, in centers and peripheries, do affect such an aim after all. They may, in fact, be necessary if an egalitarian online world is to be ensured.

Materiality and Marginalization

That a technology such as the computer microprocessor, considered a pinnacle of complex, science-led, US innovation, could wind up in a wholly unanticipated locale such as an Internet café in Accra is the current global reality of rapid, informal commodity flows and technology diffusion exemplified in many pockets of activity on the African continent. It is taking place at accelerated speed and through a multiplicity of diverse new metropoles such as Hong Kong and Dubai as the pathways through other established centers become more costly or exclusionary. This book began (and now ends) with a call for scholarship on technology development and diffusion and on the construct of "the user" to encompass the full diversity of political-economic processes. Beyond mainstream market-based technology distributions many other processes such as gray market commodity flows, the operations of the international aid sector, and global center-periphery dynamics are shaping the conditions under which technological artifacts are accessed and engaged.

This concluding chapter has focused on certain intractable and perhaps worsening asymmetries of the reproduction of global centers and peripheries in network structures as experienced by Ghanaians online. For the youth of Accra's Internet cafés, the Internet was not traveled as an individual sojourn through an impersonal information space but embraced as a social world. The effectiveness of their online forays (in the very concrete sense of improved life changes and financial gains) was dependent on the way they were received and recognized by others online, ranging from network administrators and content providers to other users. The concept of digital marginalization was offered in the introduction as a way to consider disadvantages defined relationally within a state of connectivity. I have sought to show how this marginalization experienced by young Ghanaians generated not just deprivations, but also novel engagements with the technological form, spanning a wide range of practices—from the narrative formats of Internet scamming to the manual labor of metal scrap dealing.

The extension of theories of materiality has served several purposes in the book's larger argument about the global Internet, the invisibility of this particular user population, and the concept of their digital marginalization. The material world and specifically the materiality of the Internet itself was a critical third party to new possibilities for social enactment. Studies in more mainstream threads of anthropology, sociology, and media studies, when considering technology, often fall back into the habit of depicting the symbolic play across material surfaces without delving deeply enough, I have argued, into technology's functionality and form. This book resisted this tendency through an analysis via an adapted material semiotics that provided a way to bridge between richly cultural and particularistic studies of technology adoption and notions of the global transformations of the digital age.

The consideration of materiality has also served as a counterpoint to arguments for the liberatory potential of the Internet via its immateriality, an old argument now resurfacing in some of the recent discussions about the Internet as a critical tool of socioeconomic development, given voice in particular at the World Summit on the Information Society conference (see chapter 6). The notion of cyberspace as transcending the corporeal world tends to imply a view of the technology as without limits, costs, or trade-offs and in this way hampers the possibility of critique. An alternate argument carried through these chapters is that there is no experience of immateriality that is without material aspects.[16] Virtual worlds may be thought of as spaces of reconfigured and in some ways loosened and novel materiality but they are by no means simply immaterial domains. For something as seemingly ephemeral as an idea, a belief, or an experience to be shared and thereby to be social fundamentally requires externalization (and thus materialization) of some sort. Likewise considering how the Internet is a material phenomenon is to finally contend with its limits, but it also recognizes that what users do online, in leaving behind digital traces, is often enduring and materially consequential. To acknowledge the materiality of the Internet is to see how new possibilities are concretely enacted and the firmness and force of constraints. This is one avenue for confronting the emerging asymmetries and power differentials that are carried through this new technology.

Young Ghanaian Internet users resolutely, illegitimately, with mixed success, but in some cases quite effectively, managed to redirect these foreign forms toward their own alternate purposes, their view of the world and their place within it. This punctures certain illusions of technological inevitability often propounded in the technocratic outlook of international aid efforts and the rhetoric from high-tech industry centers. These invisible users demonstrated diverse capabilities for coping with and managing a novel technological system that was not designed with them in mind. They forged a distinctive (if often thwarted) path through the vast and global network into which they were unpredictably drawn.

Notes

Chapter 1

1. The works of Miller and Slater (2000) and Horst and Miller (2006) consider diasporic zones that still maintain a considerable degree of cultural coherence.

2. I have capitalized *Traditional Culture* in order to highlight its problematic fixedness and singularity. As the anthropologist Mary Douglas notes, there is no such thing (see Douglas 2004).

3. For further allied definitions see Karin Knorr Cetina (1999) on epistemic cultures and notions of performativity in actor-network theory (Law 1999) and Karen Barad's posthumanist performativity (Barad 2003).

4. There is nothing apparently illegal about this trade, thus it is not of the black market but it is opportunistic, ad hoc, and outside of the mainstream, recognized commodity flows of the formal computing industry.

5. Examples of Africa's depiction as a technological blank slate in the rhetoric of development institutions and in relation to policy matters include a 2005 UN report titled "Open Access for Africa" that asserts in the preface that "the absence of modern tools for gathering information and communicating is particularly evident on the continent of Africa" (Danofsky 2005) and a World Bank report that positions the continent of Africa as an anachronism with the title "Can Africa Join the 21st Century?" (World Bank 2000).

6. When breakdown is an expected and routine event we can expect less urgency in action, and less of this sense of privileged analytical rarity that is reflected in *Reassembling the Social*, in which Bruno Latour identifies "accidents, breakdowns and strikes" as where the agency of objects becomes visible, "all of a sudden, completely silent intermediaries become full-blown mediators; even objects, which a minute before appeared fully automatic, autonomous, and devoid of human agents, are now made of crowds of frantically moving humans with heavy equipment" (Latour 2005, 81).

7. In social studies of science and technology a similar betrayal is couched alternately in terms of the struggle to construct knowledge in face of an unknown outcome. Laboratory work requires alliance building with material elements (Latour 1984). Yet the cultivation of an antibody may inexplicably fail. Latour (1992) amusingly describes the door closer that goes "on strike." Similarly, Gell (1998) speaks of a car that stalls and fails to restart once the driver is somewhere remote and inaccessible. In this broader context, betrayals take place not just in the way material realities are surplus to symbolic representation but also through failed attempts by users to enroll their particular properties and functions. Tenner (1996) offers a populist account of this.

8. Latour notes a relevant distinction between intermediaries and mediators. Mediators are entities that "transform, translate, destroy, and modify the meaning or the elements they are supposed to carry" whereas intermediaries "[transport] meaning or force without transformation" (Latour 2005, 39).

9. A good example of this is Latour's account of the door closer where the humans in the setting are described with no larger motives than to contend successfully with a threshold, hinges, and pneumatics that may or may not be functioning at a given time. In a brief aside Latour acknowledges that "people will start bothering about reading the maps, getting their feet muddy and pushing the door open only if they are convinced that the group is worth visiting" (Latour 1992, 240). But how to bridge between these two points? The distance appears vast.

10. See Vandenberghe (2002) for a full elaboration of this critique.

11. I refer to Marcus and Fischer's definition of political economy as "the mutual determination of political processes and economic activity in a historically viewed world system of nation states" (Marcus and Fischer 1986, 79).

12. Work by Ron Eglash (2004) represents one attempt to expand the model to accommodate these unknown user populations. He distinguishes between users with low versus high social power. Low social power includes having no (or very limited) influence over design and production processes.

13. As similarly critiqued by Suchman (2007, 192–193).

Chapter 2

1. This is a slightly broader rephrasing of a similar question, "If the Internet was the solution, what was the problem?" posed by Bakardjieva and Smith (2001, 71).

2. Weiss (2002, 98) quoting Appadurai (1991, 193).

3. *Chop* is pidgin English for *to eat.* So *chop money* in essence is one's daily food (plus transportation) allowance.

4. A term that refers to a settlement, a place where people coming from elsewhere end up living, a "stranger's place." In Accra, it was typically Muslim northerners who lived in these places referred to as *zongos*.

5. According to 2000 census data cited in the *UN Demographic Yearbook 2006*.

6. Coe (2005, 135) refers to the Akan proverb *Abofra ntumi nka mpanyinsem* (A child cannot speak the way elders do). Proverb speaking itself is something the young are generally excluded from in Ghana and the West Africa region (Newell 2000).

7. The family I stayed with in my first few weeks in Accra was headed by a young husband and father who was an early entrant to Tianshi, a multilevel marketing organization selling Chinese herbal remedies. He had switched from a career as an auto mechanic, which had not put him on track to realize the wealth he sought. He had made quite a fortune by enrolling friends and family as buyers and sellers working under him.

8. Given interruptions in schooling from lack of funds or periods of necessary employment for survival it was also common for older students to sit alongside younger ones in the same grade levels.

9. The driver's mate is the person who collects money, announces the route of the bus to those waiting at bus stops, and arranges seating. Tro-tros are the primary form of low-cost public transportation in Ghana. The vehicles and routes are not government funded or organized but rather privately owned and operated.

10. It should be noted that boarding schools do not have the automatic status they have in parts of Western Europe and the United States. Ghanaian private schools were designed on the British model around the time of independence (1957) and boarding schools are common for secondary school education in Ghana just as they were and are in Britain.

11. The program recorded from radio was broadcast in February 2005.

12. Victor Olukoju, "The Word of God on Sex and the Youth" (Lagos, Nigeria: Missions Aid International Publications, 2002).

13. This may be a reference to the fact that many Internet café users prefer to go late at night when rates are discounted. The time difference between Ghana and the United States also makes late night a good time for individuals to meet with their US chat partners.

14. BusyInternet was originally planned as the first of a chain of Internet cafés spreading through West Africa and beyond.

15. An example of this sort of alienation in terms of space and inhabiting is found in Miller's study of tenants of council housing units (government-owned properties) in London. He draws more clearly a distinction between successful and failed

appropriation by considering the varying ways people made their units more homey or personalized. These units when received "would not be regarded as an investment of their social being" and thereby had to be made more in the image of their "self-conception as households and neighbourhoods" (Miller 1988, 354). He identified an enduring state of alienation in the way these forms signified the inhabitants' status undesirably—as poor or low class or in some ways similar to people of other ethnicities they held in low regard. Many such tenants expressed a sense of futility in taking any action at all to alter their living environment.

Chapter 3

1. See also Slater and Kwami (2005) in which they contrast this characteristic use of Internet cafés with the use of mobile phones in Ghana.

2. Good examples are Facebook and other social networking sites for keeping in touch with known, "real-world" friends. Also, online dating sites are designed to introduce people with the assumption that this will progress to face-to-face meetings. Social networking sites and dating sites are popular among youth in Accra.

3. For this interpretation I draw on observations of the hesitance and resistance of foreigners to the appeals of young Ghanaians when I observed their typed reactions in chat conversations. For example, in one observed instance a scammer when asked where he was located responded "am in Ghana u?" and received a jaded, skeptical response from a foreign chat partner, "and you are looking for a husband right?"

4. In fact, Fauzia made use of precisely this expectation within peer relationships to finance her use of the Internet café, noting "sometimes I do use my own money. I save for it because I meant to use the café. Sometimes too I do go there, when I see friends I ask them to buy me the code." Of borrowing from her friends she noted, "I just say 'oh, Charley you know what I want you to buy me a code, I'm broke, you know I'm broke."

5. Several articles speak about the endurance of patron-clientalism in the West Africa and broader sub-Saharan region in the contemporary era as something that is apparent in kinship relationships as well as in government power structures (Berry 1989; Smith 2001; Bayart 2000; de Sardan 1999).

6. The term *cyberspace* was originally coined by science fiction writer William Gibson.

7. In Hausa, *sakawa* is a verb meaning *to take* or *to pick*. How it came to refer to Internet fraud as a totality is not entirely clear but a Hausa speaker I consulted suggested that it related to credit card fraud, the act of putting something into a shopping basket.

8. I received this email at random (presumably) as spam on August 15, 2006.

9. The 419 scams clearly can be successful. The annual report of the Internet Crime Complaint Center, an organization maintained by the United States Federal Bureau of Investigations and National White Collar Crime Center, shows that victims report a high median loss (at $1,650 per incident in the 2008 report [Internet Crime Complaint Center 2008]) compared to most other forms of reported fraud such as credit card fraud, delivery failure, online auction fraud, and so on.

10. These text-based game environments were referred to as MUDs (multiuser dungeons) or MOOs (MUD object-oriented).

11. Similarly in Nardi's recent ethnography of the multiplayer role-playing game *World of Warcraft* she finds gender swapping continues but is responded to very differently by Chinese versus American gamers (Nardi 2010).

12. Kitchin (1998) provides an overview of utopian visions of the transformations realized through new networking technologies. Poster (1995a, 1995b) and Turkle (1995) provide convincing and well-argued cases for this way of thinking whereas Nakamura (2002) and Robins (1995) provide critiques.

13. Approximately $5,000 at the time.

14. A pseudonym for an actual screen name.

15. As Nakamura (2002) and Robins (1995) critique.

16. The literature on online embodiment has documented this concern among users with authenticity and deception and the correlation between online representation and offline reality even in fantasy spaces that are ostensibly meant for the creation of fictional identities (see Slater 1998; Bassett 1997; Reid 1996).

17. Heavily Indebted Poor Countries (HIPC) is a World Bank initiative that has approved eighteen countries (including Ghana) for debt relief.

18. For example, an e-book titled *Information and Communication Technology for Peace* was distributed as part of the UN-sponsored World Summit on the Information Society (Stauffacher et al. 2005). In the preface, then UN secretary-general Kofi Annan suggests that "by promoting access to knowledge, [ICTs] can promote mutual understanding, an essential factor in conflict prevention and post-conflict reconciliation."

Chapter 4

1. Kapferer (1990) argues that falseness is not essential to the definition of rumor and that the stories that circulate in this way sometimes prove to be true.

2. One illustration of the weak effects attributed to speech acts is Latour's famous example of the hotel owner who wishes to prevent guests from losing hotel room keys. First the guests are instructed verbally to "bring the keys to the front desk

when you go out." Ultimately and most effectively, guests are enrolled in the key-saving program once a heavy weight attached to the key adds material force to the mere verbal request by making it inconvenient to carry around (Latour 1991).

3. It is worth noting here that these young Ghanaians thought of credit cards as bank accounts with a certain amount of money already in them. Many didn't have a notion of credit card limits or what a typical limit would be. This is additional evidence of how distant they actually were from those effectively committing Internet fraud.

4. External observers might alternately apply the term *vigilante justice,* which has more negative connotations.

5. De Witte (2001) documents the emergence of a funeral industry among the Asante in Ghana with many roles that had been taken on by the family hired out instead. My observations confirm and extend this beyond funerals to the whole range of social events in Accra and beyond the Asante to the broader range of ethnic and religious groups.

6. Kwabena Ofori Tenkorang-Mensah, *The Story of the Okyeame Network,* 2002. http://www.okyeame.net/okyeame/the_okyeame_story.html.

7. The official rules of the okyeame mailing list state "anybody who does not adhere to these guidelines will get ONE 'unfriendly' reminder from the postmaster. Subsequent violation will result in the removal of the offender from the list." See *Official Rules and Purpose of the Okyeame.Network,* 2002. http://www.okyeame.net/okyeame/rules.html.

Chapter 5

1. In this way this chapter departs from the literature that looks at religious belief as a moral code that is brought into decision making about whether to adopt a technology and how to use it (for example, see Winston 2005; Whetmore 2007; Woodruff, Augustin, and Foucault 2007).

2. Churches were the most public and prominent of these sites, but shrines, mosques, "the realm of the spirit," and certain geographic areas (regions where more powerful mallams or fetish priests worked, for example) were other sites of religious ritual relevant to Ghanaians.

3. It should be noted that an investment in this metaphysics certainly varied among Internet users and for some these forces were seen as a bit more distant and less relevant than they were to others.

4. Coastal communities have a disproportionate number of mainline Protestant church members (Presbyterian, Methodist, Anglican) because these denominations arrived with European traders, explorers, and colonists via the coast and mostly

remained there. Muslim traders came to Ghana from the north and therefore disproportionately attracted northerners to the faith. Newer Pentecostal faiths have a disproportionate number of followers in Accra; the airplane and mass media (as delivery modes that are most easily accessed in the urban center) play a role in this.

5. It is also possible to hear the Muslim call to prayers in particular neighborhoods and to witness traditional festivals such as the Ga Homowo festival at certain times of year. However, these tend to be more localized than the Christian sermons and music that are ubiquitous throughout the city and broadcast on television and radio.

6. Related to the danger of modern possessions is the story of Mami Water that is pervasive along the West Coast of Africa. This goddess who lives at the bottom of the sea is known for her fascination with and accumulation of modern electronics, autos, and jewelry. These material riches can be obtained from her oceanic stash but only through occult means and at a price. For women, the consequence of attaining material possessions through occult means is infertility (Meyer 1998a).

7. Part of Gabby's agreement with his mallam was that he would share any of the profits he gained from his online scamming.

Chapter 6

1. See Chambers (1995) for a critique of how development professionals see poverty and its solutions in ways that reflect their stance as urban, educated northerners and miss critical aspects of poverty (and potential solutions) by neglecting to consider and understand how it is alternately framed by the poor themselves.

2. Although the WSIS regional conference in Accra was for the whole of Africa, attending was likely more feasible, convenient, and affordable for participants traveling from Europe than from East Africa. This is due to limited regional air travel and the fact that transcontinental airlines generally transit through Europe. Road travel was possible for some participants within a limited radius but depended on the quality of roads and security issues.

3. Joseph Stiglitz, former chief economist of the World Bank, similarly critiques this insularity and cloistering of the IFIs where intervention work involves teams of economists pouring over numbers, isolating themselves to hotels and ultimately applying a cookie-cutter approach that is often quite ignorant of the local political or historical particularities of the nation-state in question. See Stiglitz (2000).

4. I omit from this particular criticism the domain of NGO work that (depending on the organization) often demonstrates a sustained presence in regions targeted for aid and some degree of dialog with beneficiary populations.

5. It should be noted that the notion of a multistakeholder approach was not entirely novel to World Summit proceedings. It was an element in the World Summit

on Sustainable Development in Johannesburg in 2002. However, a perhaps stronger conviction in such an approach and better procedures for ensuring inclusiveness were evident in the early preparation phase of WSIS.

6. See Rodrik (2006) for a review of this period in development aid and its decline.

7. See Annan (1999).

8. See *Global Voices: About.* http://globalvoicesonline.org/about. Posted April 9, 2007. Last updated August 26, 2011.

9. No specific statistics on participation were available for the regional conference in Accra.

10. The final list of WSIS participants at the Tunis conference is available at http://www.itu.int/wsis/documents/doc_multi.asp?lang=en&id=2294|0. No such detailed list of participants at the regional conference in Accra was available.

Chapter 7

1. As stated in the *WSIS Outcome Documents,* "The capacity of these technologies to reduce many traditional obstacles, especially those of time and distance, for the first time in history makes it possible to use the potential of these technologies for the benefit of millions of people in all corners of the world" (WSIS 2005, 10).

2. I was able to calculate this by looking at the system dialog in Windows on one of the Internet café machines to reference the computer processor make, generally an Intel Pentium III or IV. The "Intel Microprocessor Quick Reference Guide" provides the date when that particular processor came onto the market. http://www.intel.com/pressroom/kits/quickrefyr.htm.

3. Tim Mangini, "Ghana: Digital Dumping Ground," *Frontline,* June 23, 2009. http://www.pbs.org/frontlineworld/stories/ghana804/video/video_index.html.

4. Chris Caroll, "High-Tech Trash," *National Geographic,* January 2008.

5. Pieter Hugo, "A Global Graveyard for Dead Computers," *New York Times,* August 4, 2010.

6. See the 2009 report and Web site from Greenpeace International, "Where Does E-waste End Up?" (http://www.greenpeace.org/international/en/campaigns/toxics/electronics/the-e-waste-problem/where-does-e-waste-end-up) as well as the report by Puckett et al. (2005). The Basel Action Network (BAN) names itself after the Basel Convention on the Control of Transboundary Movement of Hazardous Wastes and Their Disposal, a United Nations treaty formulated in 1989 to reduce the export of waste from the developed to developing regions and that the United States has, so far, declined to sign.

7. Mangini, "Ghana: Digital Dumping Ground."

8. "Policy of Importation of Old Fridges Begins Next Year," *The Daily Graphic,* July 19, 2010.

9. See Ministry of Finance and Economic Planning of the Republic of Ghana, The Harmonized System and Customs Tariff Schedules, 2007, specifically HS heading 84.71.

10. In a conversation with a British NGO worker actively involved in the export of computers to different African countries she noted that gaining certification of nonprofit status from the government in Ghana is notably easier than in Kenya, for example. This opens up opportunities for abuse when importers claiming such status end up selling commodities. The magnitude of such abuse, however, is difficult to determine.

11. Sam Wakahakha, "Uganda to Review Ban on Import of Used Electronics," *The East African,* March 8, 2010.

12. Michael Malakata, "Uganda Reverses PC Recycling Policy as Kenya Imposes Ban," *ComputerWorld,* March 22, 2010. http://news.idg.no/cw/art.cfm?id=87B3471E-1A64-67EA-E4F713476550FBBB.

13. One computer unit is composed of a CPU (central processing unit) grouped with or without a keyboard and display (computer monitor). The relevant category in Ghanaian government's Harmonized Systems and Customs Tariff Schedules is Heading 84.71. These quantities came with the total valuation as calculated by customs. The average per-unit value for each unit was between $114 and $225 during those years. This rather low value indicates the likelihood that the bulk of these units were secondhand computers.

14. This business name is a pseudonym.

15. This figure is strikingly lower than the 75 to 80 percent waste estimate from the BAN report in Nigeria or the 50 percent from the *National Geographic* article reported from Ghana. I wish to point out that these reported figures on percentage waste in shipments of imported computers as well as the ones salesmen in Ghana reported to me are all self-reported estimates reflecting the self-interest of those doing the estimating and are highly, highly unreliable at this time.

16. In the same mode, there is a movement within Western consumer cultures toward tracing and making visible source materials and labor practices notably in the fair trade movement. Retailers such as Starbucks coffee company, which participate in this campaign, employ this visibility in marketing materials, making use of photographs depicting pastoral scenes of coffee farming and world maps labeling source countries.

17. This led to some confusing phrasing as when my Ghanaian research assistant, Kobby, asked a mobile phone merchant, "Which phones were brought from home?" and then clarified, "I mean abroad?"

18. Since the passing of Legislative Instrument (L.I.) 1586 it has been illegal to import and sell secondhand mattresses in Ghana.

19. From my recollection, the scrap collectors with wooden carts simply were not visible in this way in 2004 and 2005. My research assistant also confirmed the wooden carts as a recent phenomenon. The scrap dealers themselves noted that the business had grown and gotten more competitive since 2006 or so.

20. One team wears shirts, the other goes shirtless.

21. See United States EPA Web page, *Peristent Bioaccumulative and Toxic (PBT) Chemical Program: Dioxins and Furans.* Last modified April 18, 2011. http://www.epa.gov/pbt/pubs/dioxins.htm.

22. See United States EPA Web page, *Pollutants/Toxins > Air Pollutants > Lead.* Last modified January 20, 2011. http://www.epa.gov/ebtpages/pollairpollutantslead.html.

23. The Basel Action Network NGO, for example, has set up a program called e-stewards to certify e-waste recyclers in the United States uphold standards of waste management, such as ensuring that recyclers are not exporting any toxic waste to developing countries.

24. Mangini, "Ghana: Digital Dumping Ground."

25. Neil MacFarquhar, "Banks Making Big Profits from Tiny Loans," *New York Times*, April 13, 2010.

Chapter 8

1. The small ten-screen café named Lambos in Mamobi was the only one among my selected field sites that I found closed down on my return visit in the summer of 2010.

2. On the possibility of investing gains from scamming in something more lasting such as land, property, or a business venture, one young scammer asked me, "Can you wear a house on your head?" to the laughter of his friends. This group of young men, who from time to time attended the karaoke night, emphasized the use of these scamming gains for impressing girls, *boiling up* (a slang term for relaxing after working hard), and displaying publicly and ostentatiously the spoils of their online conquests.

3. It is useful to note that in mainstream Ghanaian society, Rastas and Rastafarian culture are often associated with violence, pot smoking, and criminal acts. The effects of pot smoking are often linked to armed robbers and thus with increased aggression and violence rather than with a lack of motivation as in Western constructions of the effects of marijuana.

4. It is a standard practice in the Nigerian (Nollywood) and Ghanaian video film industries that produces films with tiny budgets and on very short time frames to generate several sequels in rapid succession starring the same actors and following a plot that stems from the same general premise as the first film.

5. The emergence of sakawa into popular culture in Ghana is just beginning to find its way into the scholarly literature; see Meyer (2011).

6. Awudu Mahama and Sheilla Sackey, Sakawa Budget, *The Daily Guide,* March 6, 2009. http://www.modernghana.com/news/205352/1/sakawa-budget.html.

7. This resonates with similar responses in the region at other historical junctures, such as the disordering of mechanisms of wealth generation and the violation of norms of wealth redistribution in the shift to a cash economy in the colonial era. One consequence was massive wealth accumulation by certain members of the younger generation, an accumulation obtained in ways not fully understood by their elders. This trend was also accompanied by an active antiwitchcraft movement (see McLeod 1975; Parish 2002).

8. Even Gabby, who had once earnestly visited and paid several mallams (see chapter 5), said these practices didn't work and claimed he no longer visited such individuals.

9. See, for example, the two-page Verizon-Google Legislative Framework Proposal presented to the FCC on August 9, 2010 (http://static.googleusercontent.com/external_content/untrusted_dlcp/www.google.com/en/us/googleblogs/pdfs/verizon_google_legislative_framework_proposal_081010.pdf) and the Policy Statement of the Federal Communication Commission FCC 05–151 dated August 5, 2005.

10. Koranteng Ofosu-Amaah, "Black Sheep," *Koranteng's Toli* (blog), February 2, 2010. http://koranteng.blogspot.com/2010/02/black-sheep.html.

11. To locate some of this "chatter" I did a search on "ip address blocking Africa" in the main Google Search and in Google Groups to get to discussion boards.

12. See http://www.webmasterworld.com/forum22/3091-2-30.htm, last accessed April 25, 2011.

13. See http://forums.theplanet.com/lofiversion/index.php/t80479.html, last accessed April 25, 2011.

14. See http://www.cyberciti.biz/faq/block-entier-country-using-iptables, last accessed April 25, 2011.

15. See http://forums.plentyoffish.com/datingPosts5610943.aspx, last accessed April 25, 2011.

16. Miller (2005) makes this point about the inescapability of the material even in conceptualizing notions of the immaterial, as in the expression of religious faith.

References

Abbink, Jon, and Ineke van Kessel, eds. 2005. *Vanguard or Vandals: Youth, Politics and Conflict in Africa*. Leiden: Brill.

Basic Information about WSIS—Overview. 2006. http://www.itu.int/wsis/basic/about.html.

Adas, Michael. 1989. *Machines as the Measure of Men: Science, Technology, and Ideologies of Western Dominance*. Ithaca, NY: Cornell University Press.

Ahearn, Laura M. 2001. Language and Agency. *Annual Review of Anthropology* 30:109–137.

Akrich, Madeleine. 1992. The De-Scription of Technical Objects. In *Shaping Technology/Building Society: Studies in Sociotechnical Change*, ed. W. E. Bijker and J. Law. Cambridge, MA: MIT Press.

Akrich, Madeleine, and Bruno Latour. 1992. A Summary of a Convenient Vocabulary for the Semiotics of Human and Nonhuman Assemblies. In *Shaping Technology/Building Society: Studies in Sociotechnical Change*, ed. W. E. Bijker and J. Law. Cambridge, MA: MIT Press.

Anderson, Benedict. 1983. *Imagined Communities: Reflections on the Origin and Spread of Nationalism*. London: Verso.

Annan, Kofi. 1999. Secretary-General Proposes Global Compact on Human Rights, Labour, Environment, in Address to World Economic Forum in Davos. United Nations press release. http://www.un.org/News/Press/docs/1999/19990201.sgsm6881.html.

Annan, Kofi. 2002. Kofi Annan's IT Challenge to Silicon Valley. *CNet News*. http://news.cnet.com/2010-1069-964507.html.

Appadurai, Arjun. 1988. Introduction: Commodities and the Politics of Value. In *The Social Life of Things: Commodities in Cultural Perspective*, ed. A. Appadurai. Cambridge, UK: Cambridge University Press.

Appadurai, Arjun. 1991. Global Ethnoscapes: Notes and Queries for a Transnational Anthropology. In *Recapturing Anthropology: Working in the Present*, ed. Richard G. Fox, 191–210. School of American Research Press: Santa Fe, New Mexico.

Appadurai, Arjun. 1996. *Modernity at Large: Cultural Dimensions of Globalization*. Minneapolis: University of Minnesota Press.

Apter, Andrew. 1999. IBB = 419: Nigerian Democracy and the Politics of Illusion. In *Civil Society and the Political Imagination in Africa*, ed. J. L. Comaroff and J. Comaroff. Chicago: The University of Chicago Press.

Asad, Talal. 2003. *Formations of the Secular: Christianity, Islam, Modernity*. Stanford, CA: Stanford University Press.

Bakardjieva, Maria, and Richard Smith. 2001. The Internet in Everyday Life: Computing Technologies from the Standpoint of the Domestic User. *New Media & Society* 3 (1):67–84.

Barad, Karen. 2003. Posthumanist Performativity: Toward an Understanding of How Matter Comes to Matter. *Signs: Journal of Women in Culture and Society* 28 (3):801–831.

Barber, Karin. 1997. *Readings in African Popular Culture*. Bloomington: Indiana University Press.

Barlow, John Perry. 1996. *A Declaration of the Independence of Cyberspace*. Electronic Frontier Foundation. https://projects.eff.org/~barlow/Declaration-Final.html.

Bassett, Caroline. 1997. Virtually Gendered: Life in an On-line World. In *The Subcultures Reader*, ed. K. Gelder and S. Thornton. London: Routledge.

Bayart, Jean-Francois. 2000. Africa in the World: A History of Extraversion. *African Affairs* 99:217–267.

Beck, Ulrich. 1992. *Risk Society: Towards a New Modernity*. London: Sage.

Bell, Daniel. 1974. *The Coming of Post-Industrial Society: A Venture in Social Forecasting*. London: Heinemann Educational.

Benjamin, Walter. 2001 [1936]. The Work of Art in the Age of Mechanical Reproduction. In *Media and Cultural Studies: Keyworks*, ed. M. G. Durham and D. M. Kellner. Malden, MA: Blackwell.

Bernal, Victoria. 2005. Eritrea On-Line: Diaspora, Cyberspace, and the Public Sphere. *American Ethnologist* 32 (4):660–674.

Berry, Sara. 1989. Social Institutions and Access to Resources. *Africa* 59 (1):41–55.

Bijker, Wiebe E. 1995. *Of Bicycles, Bakelites, and Bulbs: Toward a Theory of Sociotechnical Change*. Cambridge, MA: MIT Press.

Bloem, Renate. 2005. Multi-Stakeholderism and Civil Society. In *The World Summit of the Information Society: Moving from the Past into the Future*, ed. D. Stauffacher and W. Kleinwächter. New York: The United Nations Information and Communication Technologies Task Force.

Boellstorff, Tom. 2010. *Coming of Age in Second Life*. Princeton, NJ: Princeton University Press.

Bourdieu, Pierre. 1977. *Outline of a Theory of Practice*. Cambridge, UK: Cambridge University Press.

Bryceson, D., and U. Vuorela. 2002. *The Transnational Family: New European Frontiers and Global Networks*. Oxford: Berg.

Brydon, Lynne, and Karen Legge. 1996. *Adjusting Society: The World Bank, the IMF, and Ghana*. London: Tauris Academic Studies.

Burke, Timothy. 1996. *Lifebuoy Men, Lux Women*. Durham, NC: Duke University Press.

Burrell, Jenna. 2008. Problematic Empowerment: West African Internet Scams as Strategic Misrepresentation. *Information Technologies and International Development* 4 (4):15–30.

Burrell, Jenna. 2009. The Fieldsite as a Network: A Strategy for Locating Ethnographic Research. *Field Methods* 21 (2):181–199.

Burrell, Jenna. 2011. User Agency in the Middle Range: Rumors and the Reinvention of the Internet in Accra, Ghana. *Science, Technology, & Human Values* 36(2):139–159.

Burrell, Jenna, and Ken Anderson. 2008. "I Have Great Desires to Look beyond My World": Trajectories of Information and Communication Technology Use among Ghanaians Living Abroad. *New Media & Society* 10 (2):203–224.

Carling, Jorgen. 2002. Migration in the Age of Involuntary Immobility: Theoretical Reflections and Cape Verdean Experiences. *Journal of Ethnic and Migration Studies* 28 (1):5–42.

Carothers, Thomas. 1999–2000. Think Again: Civil Society. *Foreign Policy* 17:18–29.

Casanova, Jose. 1994. *Public Religions in the Modern World*. Chicago: The University of Chicago Press.

Castells, Manuel. 1996. *The Rise of the Network Society*. vol. 1. Oxford: Blackwell.

Castells, Manuel. 1997. *The Power of Identity*. Oxford: Blackwell.

Castells, Manuel. 1998. *The End of the Millennium*. vol. 1. Oxford: Blackwell.

Castells, Manuel. 2001. *The Internet Galaxy: Reflections on the Internet, Business and Society*. Oxford: Oxford University Press.

Castells, Manuel, Mireia Fernandez-Ardevol, Jack Linchuan Qiu, and Araba Sey. 2006. *Mobile Communication and Society: A Global Perspective*. Cambridge, MA: MIT Press.

Chambers, Robert. 1995. Poverty and Livelihoods: Whose Reality Counts? *Environment and Urbanization* 7 (1):173–204.

Chant, Sylvia, and Gareth Jones. 2005. Youth, Gender and Livelihoods in West Africa: Perspectives from Ghana and the Gambia. *Children's Geographies* 3 (2):185–199.

Cockburn, Cynthia, and Susan Ormrod. 1993. *Gender and Technology in the Making*. London: Sage.

Coe, Cati. 2005. *Dilemmas of Culture in African Schools: Youth, Nationalism, and the Transformation of Knowledge*. Chicago: The University of Chicago Press.

Cohen, Stanley. 1972. *Folk Devils and Moral Panics: The Creation of the Mods and the Rockers*. London: Routledge.

Comaroff, Jean. 1996. The Empire's Old Clothes: Fashioning the Colonial Subject. In *Cross-Cultural Consumption: Global Markets, Local Realities*, ed. D. Howes. London: Routledge.

Comaroff, Jean, and John L. Comaroff. 1999. Occult Economies and the Violence of Abstraction: Notes from the South African Postcolony. *American Ethnologist* 26 (2):279–303.

Couldry, Nick. 2003. Passing Ethnographies: Rethinking the Sites of Agency and Reflexivity in a Mediated World. In *Global Media Studies: Ethnographic Perspectives*, ed. P. Murphy and M. Kraidy. New York: Routledge.

Cowan, Ruth Schwartz. 1987. The Consumption Junction: A Proposal for Research Strategies in the Sociology of Technology. In *The Social Construction of Technological Systems: New Directions in the Sociology and History of Technology*, ed. W. E. Bijker, T. P. Hughes, and T. J. Pinch. Cambridge, MA: MIT Press.

Cruise O'Brien, Donal. 1996. A Lost Generation? Youth Identity and State Decay in West Africa. In *Postcolonial Identities in Africa*, ed. R. P. Werbner and T. Ranger. London: Zed Books.

Danofsky, Samuel. 2005. *Open Access for Africa: Challenges, Recommendations, and Examples*. New York: United Nations ICT Task Force Working Group on the Enabling Environment.

de Bruijn, Esther. 2008. "What's Love" in an Interconnected World? Ghanaian Market Literature for Youth Responds. *Journal of Commonwealth Literature* 43 (3):3–24.

de Certeau, Michel. 1984. *The Practice of Everyday Life*. Berkeley: University of California Press.

de Laet, Marianne, and Annemarie Mol. 2000. The Zimbabwe Bush Pump: Mechanics of a Fluid Technology. *Social Studies of Science* 30 (2):225–263.

de Sardan, J. P. Olivier. 1999. A Moral Economy of Corruption in Africa? *Journal of Modern African Studies* 37:25–52.

de Witte, Marleen. 2001. *Long Live the Dead! Changing Funeral Celebrations in Asante, Ghana*. Amsterdam: Aksant.

de Witte, Marleen. 2003. Altar Media's *Living Word*: Televised Charismatic Christianity in Ghana. *Journal of Religion in Africa* 33 (2):172–202.

Deibert, Ronald, John Palfrey, Rafal Rohozinski, and Jonathan Zittrain, eds. 2008. *Access Denied: The Practice and Policy of Global Internet Filtering*. Cambridge, MA: MIT Press.

Donner, Jonathan. 2007. The Rules of Beeping: Exchanging Messages via Intentional "Missed Calls" on Mobile Phones. *Journal of Computer-Mediated Communication* 13 (1):37–51.

Douglas, Mary. 2004. Traditional Culture—Let's Hear No More about It. In *Culture and Public Action*, ed. V. Rao and M. Walton. Stanford, CA: Stanford University Press.

Douglas, Mary, and Baron Isherwood. 1979. *The World of Goods*. New York: Basic Books.

Du Bois, W. E. B. [1903] 2003. *The Souls of Black Folk*. New York: Barnes and Noble Classics.

Durkheim, Emile. [1912] 1995. *The Elementary Forms of Religious Life*. New York: Free Press.

Easterly, William. 2006. *The White Man's Burden: Why the West's Efforts to Aid the Rest Have Done so Much Ill and so Little Good*. New York: Penguin.

Ebo, Bosah, ed. 2001. *Cyberimperialism? Global Relations in the New Electronic Frontier*. Westport, CN: Praeger.

Ebron, Paulla A. 2002. *Performing Africa*. Princeton, NJ: Princeton University Press.

Eglash, Ron. 2004. Appropriating Technology: An Introduction. In *Appropriating Technology: Vernacular Science and Social Power*, ed. R. Eglash, J. Croissant, G. D. Chiro, and R. Fouche. Minneapolis: University of Minnesota Press.

Eickelman, Dale F., and Armando Salvatore. 2002. The Public Sphere and Muslim Identities. *European Journal of Sociology* 43:92–115.

Elyachar, Julia. 2002. Empowerment Money: The World Bank, Non-Governmental Organizations, and the Value of Culture in Egypt. *Public Culture* 14 (3):493–513.

Englund, Harri. 2004. Cosmopolitanism and the Devil in Malawi. *Ethnos* 69 (3):293–316.

Escobar, Arturo. 1992. Reflections on "Development": Grassroots Approaches and Alternative Politics in the Third World. *Futures* 24 (5):411–436.

Ess, Charles. 1996. The Political Computer: Democracy, CMC, and Habermas. In *Philosophical Perspectives on Computer-Mediated Communication*, ed. C. Ess. Albany: SUNY Press.

Feldman-Savelsberg, Pamela, Flavien T. Ndonko, and Bergis Schmidt-Ehry. 2000. Sterilizing Vaccines or the Politics of the Womb: Retrospective Study of a Rumor in Cameroon. *Medical Anthropology Quarterly* 14 (2):159–179.

Ferguson, James. 1992. The Country and the City on the Copperbelt. *Cultural Anthropology* 7 (1):80–92.

Ferguson, James. 1999. *Expectations of Modernity: Myths and Meanings of Urban Life on the Zambian Copperbelt*. Berkeley: University of California Press.

Ferguson, James. 2006. *Global Shadows: Africa in the Neoliberal World Order*. Durham, NC: Duke University Press.

Fife, Elizabeth, and Laura Hosman. 2007. Public Private Partnerships and the Prospects for Sustainable ICT Projects in the Developing World. *Journal of Business Systems, Governance and Ethics* 2 (3):53–66.

Finnegan, Ruth. 2007. *The Oral and Beyond: How to Do Things with Words in Africa*. Oxford: James Currey.

Fischer, Claude. 1992. *America Calling: A Social History of the Telephone to 1940*. Berkeley: University of California Press.

Fisher, William. 1997. Doing Good? The Politics and Antipolitics of NGO Practices. *Annual Review of Anthropology* 26:39–64.

Foster, William, Seymour Goodman, Eric Osiakwan, and Adam Bernstein. 2004. Global Diffusion of the Internet IV: The Internet in Ghana. *Communications of the Association for Information Systems* 13 (38):1–47.

Garfinkel, Harold. 1967. *Studies in Ethnomethodology*. Cambridge, UK: Polity Press.

Gell, Alfred. 1988. Technology and Magic. *Anthropology Today* 4 (2):6–9.

Gell, Alfred. 1998. *Art and Agency: An Anthropological Theory*. Oxford: Oxford University Press.

Gifford, Paul. 1990. Prosperity: A New and Foreign Element in African Christianity. *Religion* 20:373–388.

Gifford, Paul. 2004. *Ghana's New Christianity: Pentecostalism in a Globalizing African Economy*. Bloomington: Indiana University Press.

Ginsburg, Faye. 1994. Embedded Aesthetics: Creating a Discursive Space for Indigenous Media. *Cultural Anthropology* 9 (3):365–382.

Ginsburg, Faye D., Lila Abu-Lughod, and Brian Larkin, eds. 2002. *Media Worlds: Anthropology on New Terrain*. Berkeley: University of California Press.

Glickman, Harvey. 2005. The Nigerian "419" Advance Fee Scams: Prank or Peril? *Canadian Journal of African Studies* 39 (3):460.

Goffman, Erving. 1996 [1959]. *The Presentation of Self in Everyday Life*. New York: Doubleday.

Goody, Esther N., and Christine Muir Groothues. 1977. The West Africans: The Quest for Education. In *Between Two Cultures: Migrants and Minorities in Britain*, ed. J. Watson. Oxford: Blackwell.

Green, Nicola. 1999. Disrupting the Field: Virtual Reality Technologies and "Multi-sited" Ethnographic Methods. *American Behavioral Scientist* 43 (3):409–421.

Guarnizo, Luis Eduardo, Alejandro Portes, and William Haller. 2003. Assimilation and Transnationalism: Determinants of Transnational Political Action among Contemporary Migrants. *American Journal of Sociology* 108 (6):1211–1248.

Gueye, C. 2003. New Information and Communication Technology Use by Muslim Mourides in Senegal. *Review of African Political Economy* 98:609–625.

Gupta, Akhil. 2000. *Postcolonial Developments: Agriculture in the Making of Modern India*. Durham, NC: Duke University Press.

Habermas, Jürgen. [1962] 1991. *The Structural Transformation of the Public Sphere: An Inquiry into a Category of Bourgeois Society*. Trans. T. Burger. Cambridge, MA: MIT Press.

Hackett, Rosalind I. J. 1995. The Gospel of Prosperity in West Africa. In *Religion and the Transformations of Capitalism: Comparative Approaches*, ed. R. H. Roberts. London: Routledge.

Hackett, Rosalind. 2005. Rethinking the Role of Religion in Changing Public Spheres: Some Comparative Perspectives. Brigham Young University Law Review 3: 659–682.

Hannerz, Ulf. 1987. The World in Creolisation. *Africa* 57 (4):546–559.

Hannerz, Ulf. 1990. Cosmopolitans and Locals in World Culture. *Theory, Culture & Society* 7 (2):237–251.

Hannerz, Ulf. 1992. *Cultural Complexity: Studies in the Social Organization of Meaning*. New York: Columbia University Press.

Hayles, N. Katherine. 1999. *How We Became Posthuman: Virtual Bodies in Cybernetics, Literature, and Informatics*. Chicago: The University of Chicago Press.

Headrick, Daniel R. 1981. *The Tools of Empire: Technology and European Imperialism in the Nineteenth Century*. Oxford: Oxford University Press.

Helmore, Edward, and Robin McKie. 2000, November 5. Gates Loses Faith in Computers: They Can't Cure World's Ills, Admits Microsoft Boss. *The Observer*, 5.

Herman, Ed, and Robert Waterman McChesney. 1997. *The Global Media: The Missionaries of Global Capitalism*. London: Cassell.

Herring, Susan C. 1993. Gender and Democracy in Computer-Mediated Communication. *Electronic Journal of Communication* 3 (2). http://www.cios.org/EJCPUBLIC/003/2/00328.HTML.

Hill, Polly. 1986. *Development Economics on Trial*. Cambridge, UK: Cambridge University Press.

Hine, Christine. 2000. *Virtual Ethnography*. London: Sage.

Hoover, Stewart M., and Lynn Schofield Clark, eds. 2002. *Practicing Religion in the Age of the Media: Explorations in Media, Religion, and Culture*. New York: Columbia University Press.

Horst, Heather, and Daniel Miller. 2006. *The Cell Phone: An Anthropology of Communication*. vol. 6. London: Berg.

Horton, Robin. 1971. African Conversion. *Africa* 41 (2):85–108.

Hubak, Marit. 1996. The Car as a Cultural Statement: Car Advertising as Gendered Socio-technical Scripts. In *Making Technology Our Own? Domesticating Technology in Everyday Life*, ed. M. Lie and K. H. Sorensen. Oslo: Scandinavian University Press.

Internet Crime Complaint Center. 2008. Internet Crime Report. http://www.ic3.gov/media/annualreport/2008_ic3report.pdf.

Jasanoff, Sheila. 2002. New Modernities: Reimagining Science, Technology, and Development. *Environmental Values* 11 (3):253–276.

Jua, Nantang. 2003. Differential Responses to Disappearing Transitional Pathways: Redefining Possibilities among Cameroonian Youth. *African Studies Review* 46 (2):13–36.

Kalathil, Shanthi, and Taylor Boas. 2001. The Internet and State Control in Authoritarian Regimes: China, Cuba, and the Counterrevolution. *First Monday* 6 (8). http://firstmonday.org/htbin/cgiwrap/bin/ojs/index.php/fm/article/view/1788.

Kapferer, Jean-Noël. 1990. *Rumors: Uses, Interpretations, and Images*. New Brunswick, NJ: Transaction.

Kelman, Ari. 2009. Even Paranoids Have Enemies: Rumors of Levee Sabotage in New Orleans' Lower 9th Ward. *Journal of Urban History* 35 (5):627–639.

Kitchin, Rob. 1998. *Cyberspace: The World in the Wires*. Chichester: Wiley.

Kivisto, Peter. 2003. Social Spaces, Transnational Immigrant Communities, and the Politics of Incorporation. *Ethnicities* 3 (1):5–28.

Kline, Ronald, and Trevor Pinch. 1996. Users as Agents of Technological Change: The Social Construction of the Automobile in the Rural United States. *Technology and Culture* 37 (4):763–795.

Knorr Cetina, Karin. 1999. *Epistemic Cultures: How the Sciences Make Knowledge*. Cambridge, MA: Harvard University Press.

Kuriyan, Renee, and Isha Ray. 2009. Outsourcing the State? Public-Private Partnerships and Information Technologies in India. *World Development* 37 (10):1663–1673.

Ladou, Joseph, and Sandra Lovegrove. 2008. Export of Electronics Equipment Waste. *International Journal of Occupational and Environmental Health* 14:1–10.

Laegran, Anne Sofie, and James Stewart. 2003. Nerdy, Trendy or Healthy? Configuring the Internet Café. *New Media & Society* 5 (3):357–377.

Lally, Elaine. 2002. *At Home with Computers*. Oxford: Berg.

Lanier, Jared, and Frank Biocca. 1992. An Insider's View of the Future of Virtual Reality. *Journal of Communication* 42(4):150–172.

Larkin, Brian. 2002. The Materiality of Cinema Theaters in Northern Nigeria. In *Media Worlds: Anthropology on New Terrain*, ed. F. D. Ginsburg, L. Abu-Lughod, and B. Larkin. Berkeley: University of California Press.

Larkin, Brian. 2008. *Signal and Noise: Media, Infrastructure and Urban Culture in Nigeria*. Durham, NC: Duke University Press.

Latour, Bruno. 1984. *The Pasteurization of France*. Cambridge, MA: Harvard University Press.

Latour, Bruno. 1987. *Science in Action: How to Follow Scientists and Engineers through Society*. Cambridge, MA: Harvard University Press.

Latour, Bruno. 1991. Technology Is Society Made Durable. In *A Sociology of Monsters: Essays on Power, Technology and Domination*, ed. J. Law. London: Routledge.

Latour, Bruno. 1992. Where Are the Missing Masses? The Sociology of a Few Mundane Artifacts. In *Shaping Technology/Building Society: Studies in Sociotechnical Change*, ed. W. E. Bijker and J. Law. Cambridge, MA: MIT Press.

Latour, Bruno. 1993. *We Have Never Been Modern*. New York: Harvester Wheatsheaf.

Latour, Bruno. 2002. Morality and Technology: The End of the Means. *Theory, Culture & Society* 19 (5/6):247–260.

Latour, Bruno. 2005. *Reassembling the Social: An Introduction to Actor-Network-Theory*. Oxford: Oxford University Press.

Law, John. 1994. *Organizing Modernity*. Oxford: Blackwell.

Law, John. 1999. After ANT: Complexity, Naming and Topology. In *Actor Network Theory and After*, ed. J. Law and J. Hassard. Oxford: Blackwell.

Law, John. 2002. Objects and Spaces. *Theory, Culture & Society* 19 (5/6):91–105.

Lea, Martin, Tim O'Shea, Pat Fung, and Russell Spears. 1992. "Flaming" in Computer-mediated Communication: Observations, Explanations, Implications. In *Contexts of Computer-Mediated Communication*, ed. M. Lea. New York: Harvester Wheatsheaf.

Lee, Benjamin, and Edward LiPuma. 2002. Cultures of Circulation: The Imaginations of Modernity. *Public Culture* 14 (1):191–213.

Lessig, Lawrence. 2001. The Internet under Siege. *Foreign Policy* 127:56–65.

Lessig, Lawrence. 2006. *Code v2*. New York: Basic Books.

Levitt, Peggy. 2001. *The Transnational Villagers*. Berkeley: University of California Press.

Livingstone, Sonia. 2002. *Young People and New Media*. London: Sage.

Lubkemann, Stephen. 2008. Involuntary Immobility: On a Theoretical Invisibility in Forced Migration Studies. *Journal of Refugee Studies* 21 (4):454–475.

Ludlow, Peter. 1996. *High Noon on the Electronic Frontier: Conceptual Issues in Cyberspace*. Cambridge, MA: MIT Press.

Mackay, Hughie, and Gareth Gillespie. 1992. Extending the Social Shaping of Technology Approach: Ideology and Appropriation. *Social Studies of Science* 22 (4):685–716.

Malinowski, Bronislaw. [1935] 1978. *Coral Gardens and Their Magic*. Mineola, NY: Dover.

Marcus, George. 1998. Ethnography in/of the World System: The Emergence of Multi-sited Ethnography. In *Ethnography through Thick and Thin*, ed. G. Marcus. Princeton, NJ: Princeton University Press.

Marcus, George E., and Michael M. J. Fischer. 1986. *Anthropology as Cultural Critique: An Experimental Moment in the Human Sciences*. Chicago: The University of Chicago Press.

Marvin, Carolyn. 1988. *When Old Technologies Were New: Thinking about Electric Communication in the Late 19th Century*. New York: Oxford University Press.

Mattelart, Armand. 1985. *Transnationals and the Third World: The Struggle for Culture*. New York: Bergin & Garvey.

Mauss, Marcel. [1950] 1990. *The Gift: The Form and Reason for Exchange in Archaic Societies*. New York: W. W. Norton.

Mbembe, Achille. 2001. *On the Postcolony*. Berkeley: University of California Press.

McLeod, Malcolm. 1975. On the Spread of Anti-Witchcraft Cults in Modern Asante. In *Changing Social Structure in Ghana: Essays in the Comparative Sociology of a New State and an Old Tradition*, ed. J. Goody. London: International African Institute.

McLuhan, Marshall. 1964. *Understanding Media: The Extensions of Man*. New York: McGraw-Hill.

McRae, Shannon. 1997. Flesh Made Word: Sex, Text and the Virtual Body. In *Internet Culture*, ed. D. Porter. New York: Routledge.

Mercer, Claire. 2005. Telecentres and Transformations: Modernizing Tanzania Through the Internet. *African Affairs* 105(419):243–264.

Meyer, Birgit. 1995. "Delivered from the Powers of Darkness" Confessions of Satanic Riches in Christian Ghana. *Africa* 65 (2):236–255.

Meyer, Birgit. 1998a. Commodities and the Power of Prayer: Pentecostalist Attitudes towards Consumption in Contemporary Ghana. *Development and Change* 29 (4):751–776.

Meyer, Birgit. 1998b. "Make a Complete Break with the Past." Memory and Post-colonial Modernity in Ghanaian Pentecostalist Discourse. *Journal of Religion in Africa* 28 (3):316–349.

Meyer, Birgit. 1998c. The Power of Money: Politics, Occult Forces, and Pentecostalism in Ghana. *African Studies Review* 41 (3):15–37.

Meyer, Birgit. 2006. Impossible Representations: Pentecostalism, Vision, and Video Technology in Ghana. In *Religion, Media, and the Public Sphere*, ed. B. Meyer and A. Moors. Bloomington: Indiana University Press.

Meyer, Birgit. 2011. Mediation and Immediacy: Sensational Forms, Semiotic Ideologies and the Question of the Medium. *Social Anthropology* 19 (1):23–39.

Meyer, Birgit, and Annelies Moors. 2006. Introduction. In *Religion, Media, and the Public Sphere*, ed. B. Meyer and A. Moors. Bloomington: Indiana University Press.

Miller, Daniel. 1987. *Material Culture and Mass Consumption*. Oxford: Basil Blackwell.

Miller, Daniel. 1988. Appropriating the State on the Council Estate. *Man* 23 (2):353–372.

Miller, Daniel, ed. 2005. *Materiality*. Durham, NC: Duke University Press.

Miller, Daniel, and Don Slater. 2000. *The Internet: An Ethnographic Approach*. London: Berg.

Mitchell, William. 1996. *City of Bits: Space, Place, and the Infobahn*. Cambridge, MA: MIT Press.

Mohan, Giles. 2000. Participatory Development and Empowerment: The Dangers of Localism. *Third World Quarterly* 21 (2):247–268.

Nakamura, Lisa. 2002. *Cybertypes: Race, Ethnicity, and Identity on the Internet*. New York: Routledge.

Nardi, Bonnie A. 2010. *My Life as a Night Elf Priest: An Anthropological Account of World of Warcraft*. Ann Arbor: University of Michigan Press.

Negroponte, Nicholas. 1995. *Being Digital*. New York: Alfred A. Knopf.

Newell, Stephanie. 2000. *Ghanaian Popular Fiction: "Thrilling Discoveries in Conjugal Life" & Other Tales*. Oxford: James Currey.

Nyamnjoh, Francis, and Ben Page. 2002. Whiteman Kontri and the Enduring Allure of Modernity among Cameroonian Youth. *African Affairs* 101:607–634.

Nye, David. 1996. *American Technological Sublime*. Cambridge, MA: MIT Press.

Odedra, M., M. Lawrie, M. Bennett, and S. Goodman. 1993. Sub-Saharan Africa: A Technological Desert. *Communications of the ACM* 36 (2):25–29.

Oldenburg, Ray. 1999. *The Great Good Place: Cafés, Coffee Shops, Bookstores, Bars, Hair Salons, and Other Hangouts at the Heart of a Community*. New York: Marlowe & Company.

Olukoju, Victor. 2002. *The Word of God on Sex and the Youth*. Lagos, Nigeria: Missions Aid International Publications.

Ong, Aihwa, and Donald Nonini, eds. 1997. *Ungrounded Empires: The Cultural Politics of Modern Chinese Transnationalism*. New York: Routledge.

Ott, Dana. 2001. Tipping the Scales? The Influence of the Internet on State-Society Relations in Africa. *Mots Pluriels* 18. http://www.arts.uwa.edu.au/MotsPluriels/MP1801do.html.

Ott, Dana, and Melissa Rosser. 2000. The Electronic Republic? The Role of the Internet in Promoting Democracy in Africa. *Democratization* 7 (1):137–156.

Oudshoorn, Nelly, and Trevor Pinch. 2008. User-Technology Relationships: Some Recent Developments. In *The Handbook of Science and Technology Studies*, ed.

E. Hackett, O. Amsterdamska, M. Lynch, and J. Wajcman. Cambridge, MA: MIT Press.

Paine, Robert. 1967. What Is Gossip About? An Alternative Hypothesis. *Man* 2 (2):278–285.

Parish, Jane. 2002. Black Market, Free Market: Anti-witchcraft Shrines and Fetishes among the Akan. In *Magical Interpretations, Material Realities: Modernity, Witchcraft, and the Occult in Postcolonial Africa*, ed. H. L. Moore and T. Sanders. London: Routledge.

Parish, Jane. 2003. Anti-witchcraft Shrines among the Akan: Possession and the Gathering of Knowledge. *African Studies Review* 46 (3):17–34.

Parish, Jane. 2011. West African Witchcraft, Wealth, and Moral Decay in New York City. *Ethnography* 12(2): 247–265.

Peha, Jon M. 2007. The Benefits and Risks of Mandating Network Neutrality, and the Quest for a Balanced Policy. *International Journal of Communication* 1:644–668.

Peil, Margaret. 1995. Ghanaians Abroad. *African Affairs* 94:345–367.

Piot, Charles. 1999. *Remotely Global: Village Modernity in West Africa*. Chicago: The University of Chicago Press.

Porat, Marc. 1977. *The Information Economy: Definition and Measurement*. Washington, DC: US Department of Commerce.

Poster, Mark. 1995a. *CyberDemocracy: Internet and the Public Sphere*. http://www.hnet.uci.edu/mposter/writings/democ.html.

Poster, Mark. 1995b. Postmodern Virtualities. In *Cyberspace, Cyberbodies, Cyberpunk: Cultures of Technological Embodiment*, ed. M. Featherstone and R. Burrows. London: Routledge.

Prins, Harald E. L. 2002. Visual Media and the Primitivist Perplex: Colonial Fantasies, Indigenous Imagination, and Advocacy in North America. In *Media Worlds: Anthropology on New Terrain*, ed. F. D. Ginsburg, L. Abu-Lughod, and B. Larkin. Berkeley: University of California Press.

Puckett, Jim, Sarah Westervelt, Richard Gutierrez, and Yuka Takamiya. 2005. *The Digital Dump: Exporting Re-use and Abuse to Africa: The Basel Action Network*. Seattle: Basel Action Network.

Raboy, Marc. 2004. The Origins of Civil Society Involvement in the WSIS. *Information Technologies and International Development* 1 (3–4):95–96.

Rahman, Aminur. 1999. *Women and Microcredit in Rural Bangladesh: An Anthropological Study of Grameen Bank Lending*. Boulder, CO: Westview Press.

Reich, Robert. 1998. The New Meaning of Corporate Social Responsibility. *California Management Review* 40 (2):8–17.

Reid, Elizabeth. 1996. Text-Based Virtual Realities: Identity and the Cyborg Body. In *High Noon on the Electronic Frontier: Conceptual Issues in Cyberspace*, ed. P. Ludlow. Cambridge, MA: MIT Press.

Rheingold, Howard. 1993. *The Virtual Community: Homesteading on the Electronic Frontier*. Reading, MA: Addison-Wesley.

Riles, Annelise. 2001. *The Network Inside Out*. Ann Arbor: University of Michigan Press.

Roberts, Lynne D. and Malcolm R. Parks. 2001. The Social Geography of Gender-Swapping in Virtual Environments on the Internet. In *Virtual Gender: Technology, Consumption, and Identity*, ed. E. Green and A. Adam. London: Routledge.

Robins, Kevin. 1995. Cyberspace and the World We Live In. In *Cyberspace, Cyberbodies, Cyberpunk: Cultures of Technological Embodiment*, ed. M. Featherstone and R. Burrows. London: Routledge.

Robins, Kevin, and Frank Webster. 1999. *Times of the Technoculture: From the Information Society to the Virtual Life*. London: Routledge.

Rodrik, Dani. 2006. Goodbye Washington Consensus, Hello Washington Confusion? A Review of the World Bank's Economic Growth in the 1990s: Learning from a Decade of Reform. *Journal of Economic Literature* 44 (4):973–987.

Rogers, Everett. 1995. *Diffusion of Innovations*. New York: Free Press.

Roman, Raul, and Royal D. Colle. 2002. *Themes and Issues in Telecentre Sustainability*. Manchester: Institute for Development Policy and Management.

Said, Edward. 1978. *Orientalism*. New York: Vintage.

Scholte, Jan Aart. 2002. Civil Society and Democracy in Global Governance. *Global Governance* 8:281–304.

Scott, James C. 1998. *Seeing Like a State: How Certain Schemes to Improve the Human Condition Have Failed*. New Haven: Yale University Press.

Sherman, Barrie, and Phil Judkins. 1992. *Glimpses of Heaven, Visions of Hell: Virtual Reality and Its Implications*. London: Hodder & Stoughton.

Shibutani, Tamotsu. 1966. *Improvised News: A Sociological Study of Rumor*. Indianapolis: Bobbs-Merrill.

Shipley, Jesse Weaver. 2009. Aesthetic of the Entrepreneur: Afro-Cosmopolitan Rap and Moral Circulation in Accra, Ghana. *Anthropological Quarterly* 82 (3):631–668.

Shiva, Vandana. 1991. The Green Revolution in Punjab. *Ecologist* 21 (2): 57–60.

Shove, Elizabeth. 2003. *Comfort, Cleanliness, and Convenience: The Social Organization of Normality*. Oxford: Berg.

Silverstone, Roger, Eric Hirsch, and David Morley. 1992. Information and Communication Technologies and the Moral Economy of the Household. In *Consuming Technologies: Media and Information in Domestic Spaces*, ed. R. Silverstone and E. Hirsch. London: Routledge.

Slade, Giles. 2006. *Made to Break: Technology and Obsolescence in America*. Cambridge, MA: Harvard University Press.

Slater, Don. 1997. *Consumer Culture and Modernity*. Cambridge, UK: Polity Press.

Slater, Don. 1998. Trading Sexpics on IRC: Embodiment and Authenticity on the Internet. *Body & Society* 4 (4):91–117.

Slater, Don, and Janet Kwami. 2005. Embeddedness and Escape: Internet and Mobile Use as Poverty Reduction Strategies in Ghana. In Working Paper Series. London: Information Society Research Group.

Smillie, Ian. 2006. *Mastering the Machine Revisited: Poverty, Aid and Technology*. Warwickshire: Practical Action Publishing.

Smith, Daniel Jordan. 2001. Ritual Killing, 419, and Fast Wealth: Inequality and the Popular Imagination in Southeastern Nigeria. *American Ethnologist* 28 (4):803–826.

Smith, Daniel Jordan. 2007. *A Culture of Corruption: Everyday Deception and Popular Discontent in Nigeria*. Princeton, NJ: Princeton University Press.

Smith, Michael Peter, and Luis Guarnizo, eds. 2002. *Transnationalism from Below*. New Brunswick, NJ: Transaction.

Spitulnik, Debra. 2002a. Alternative Small Media and Communicative Spaces. In *Media and Democracy in Africa*, ed. G. Hyden, M. Leslie, and F. Ogundimu. New Brunswick, NJ: Transaction.

Spitulnik, Debra. 2002b. Mobile Machines and Fluid Audiences: Rethinking Reception through Zambian Radio Culture. In *Media Worlds: Anthropology on New Terrain*, ed. F. D. Ginsburg, L. Abu-Lughod, and B. Larkin. Berkeley: University of California Press.

Stauffacher, Daniel, William Drake, Paul Currion, and Julia Steinberger. 2005. *Information and Communication Technology for Peace: The Role of ICT in Preventing, Responding to, and Recovering from Conflict*. New York: United Nations ICT Task Force.

Stauffacher, Daniel, and Wolfgang Kleinwächter. 2005. *The World Summit of the Information Society: Moving from the Past into the Future*. New York: The United Nations Information and Communication Technologies Task Force.

Stiglitz, Joseph. 2000. The Insider—What I Learned at the World Economic Crisis. *The New Republic*, 56.

Suchman, Lucy. 2007. *Human-Machine Reconfigurations: Plans and Situated Actions*. 2nd ed. Cambridge, UK: Cambridge University Press.

Sundén, Jenny. 2002. Cyberbodies: Writing Gender in Digital Self-Presentations. In *Digital Borderlands: Cultural Studies of Identity and Interactivity on the Internet*, ed. J. Fornäs, K. Klein, M. Ladendorf, J. Sundén, and M. Sveningsson. New York: Peter Lang.

Sundén, Jenny. 2003. *Material Virtualities: Approaching Online Textual Embodiment*. New York: Peter Lang.

Taussig, Michael. 1993. *Mimesis and Alterity: A Particular History of the Senses*. New York: Routledge.

Taylor, Charles. 2002. Modern Social Imaginaries. *Public Culture* 14 (1):91–124.

Taylor, T. L. 1999. Life in Virtual Worlds: Plural Existence, Multimodalities, and Other Online Research Challenges. *American Behavioral Scientist* 43 (3):436–449.

Tenner, Edward. 1996. *Why Things Bit Back*. New York: Alfred A. Knopf.

Turkle, Sherry. 1995. *Life on the Screen: Identity in the Age of the Internet*. New York: Simon and Schuster.

Turner, Fred. 2006. *From Counterculture to Cyberculture: Stewart Brand, the Whole Earth Network, and the Rise of Digital Utopianism*. Chicago: The University of Chicago Press.

Turner, Patricia A. 1993. *I Heard It through the Grapevine: Rumor in African-American Culture*. Berkeley: University of California Press.

Turner, Terence. 1992. Defiant Images: The Kayapo Appropriation of Video. *Anthropology Today* 8 (6):5–16.

United Nations Department of Public Information. 1999. United Nations Conferences: What Have They Accomplished? http://www.un.org/News/facts/confercs.htm [no longer available].

United Nations. 2002. Resolution adopted by the General Assembly: 56/183. World Summit on the Information Society. http://www.itu.int/wsis/docs/background/resolutions/56_183_unga_2002.pdf.

UN Demographic Yearbook. 2006. Department of Economic and Social Affairs. United Nations.

Vandenberghe, Frederic. 2002. Reconstructing Humants: A Humanist Critique of Actant-Network Theory. *Theory, Culture & Society* 19 (5/6):51–67.

Verrips, Jojada, and Birgit Meyer. 2001. Kwaku's Car: The Struggles and Stories of a Ghanaian Long-Distance Taxi-Driver. In *Car Cultures*, ed. D. Miller. Oxford: Berg.

Wakeford, Nina. 1999. Gender and the Landscapes of Computing in an Internet Café. In *Virtual Geographies: Bodies, Spaces and Relations*, ed. M. Crang, P. Crang, and J. May. London: Routledge.

Wakeford, Nina. 2003. The Embedding of Local Culture in Global Communication: Independent Internet Cafés in London. *New Media & Society* 5 (3):379–399.

Waldinger, Roger, and David Fitzgerald. 2004. Transnationalism in Question. *American Journal of Sociology* 109 (5):1177–1195.

Webster, Frank. 2002. The Information Society Revisited. In *The Handbook of New Media*, ed. L. Lievrouw and S. Livingstone. London: Sage.

Weiss, Brad. 2002. Thug Realism: Inhabiting Fantasy in Urban Tanzania. *Cultural Anthropology* 17 (1):93–124.

Whetmore, J. M. 2007. Amish Technology: Reinforcing Values and Building Community. *Technology and Society* 26 (2):10–21.

White, Luise. 2000. *Speaking with Vampires: Rumor and History in Colonial Africa*. Berkeley: University of California Press.

Willetts, Peter. 2006. The Cardoso Report on the UN and Civil Society: Functionalism, Global Corporatism, or Global Democracy? *Global Governance* 12:305–324.

Willis, Paul. 1990. *Common Culture: Symbolic Work at Play in the Everyday Cultures of the Young*. Milton Keynes, UK: Open University Press.

Wilson, Ernest J. 2006. *The Information Revolution and Developing Countries*. Cambridge, MA: MIT Press.

Winston, Hella. 2005. *Unchosen: The Hidden Lives of Hasidic Rebels*. Boston: Beacon Press.

Woodruff, Allison, Sally Augustin, and Brooke Foucault. 2007. Sabbath Day Home Automation: "It's Like Mixing Technology and Religion." In *Proceedings of the SIGCHI Conference on Human Factors in Computing Systems*. San Jose, CA: ACM.

Woolgar, Steve. 1991. Configuring the User: The Case of Usability Trials. In *A Sociology of Monsters: Essays on Power, Technology and Domination*, ed. J. Law. London: Routledge.

World Bank. 2000. Can Africa Claim the 21st Century? Washington, DC. http://www.worldbank.org/html/extdr/canafricaclaim.pdf.

WSIS. 2003. Accreditation of NGOs, Civil Society, and Business Sector Entities to the World Summit on the Information Society. Document WSIS/PC-3/7-E. World

Summit on the Information Society. http://www.itu.int/dms_pub/itu-s/md/03/wsispc3/doc/S03-WSISPC3-DOC-0007!!PDF-E.pdf.

WSIS. 2005.*WSIS Outcome Documents*. Geneva: International Telecommunication Union.

Yankah, Kwesi. 1995. *Speaking for the Chief: Okyeame and the Politics of Akan Royal Oratory*. Bloomington: Indiana University Press.

Index